Revision

The School Mathematics Project

Edited by   Stan Dolan
Barrie Hunt
Alan Knighton
Thelma Wilson
Phil Wood

Project director   Stan Dolan

The draft edition of this book was edited by Barrie Hunt. Many others helped with contributions and criticism.

The authors would like to give special thanks to Ann White for her help in preparing this book for publication.

PUBLISHED BY THE PRESS SYNDICATE OF THE UNIVERSITY OF CAMBRIDGE
The Pitt Building, Trumpington Street, Cambridge, United Kingdom

CAMBRIDGE UNIVERSITY PRESS
The Edinburgh Building, Cambridge CB2 2RU, UK
40 West 20th Street, New York, NY 10011–4211, USA
10 Stamford Road, Oakleigh, VIC 3166, Australia
Ruiz de Alarcón 13, 28014 Madrid, Spain
Dock House, The Waterfront, Cape Town 8001, South Africa

http://www.cambridge.org

First published 1992
Seventh printing 2000

Produced by 16–19 Mathematics

Printed in the United Kingdom at the University Press, Cambridge

ISBN 0 521 42394 5

Contents

FOUNDATIONS

Straight lines   The equation of a straight-line graph is

$$y = mx + c$$

where m is the gradient and c is the intercept on the y-axis.

The line through (a, b) and (c, d) has equation

$$y - b = \frac{d-b}{c-a}(x - a)$$

Quadratic functions   A quadratic function in expanded form, such as $x^2 + 6x + 5$

may also be expressed in factorised form $(x + 1)(x + 5)$

or in completed square form $(x + 3)^2 - 4$

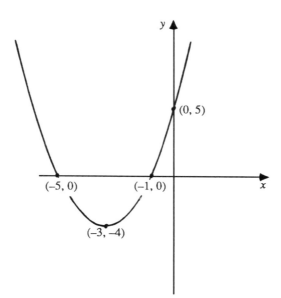

The intercept $(0, 5)$ is evident from the expanded form

$$y = x^2 + 6x + 5$$

The points $(-1, 0)$ and $(-5, 0)$ are evident from the factorised form

$$y = (x + 1)(x + 5)$$

The vertex $(-3, -4)$ can be obtained easily from the completed square form

$$y = (x + 3)^2 - 4$$

The graph is a translation of $y = x^2$ through $\begin{bmatrix} -3 \\ -4 \end{bmatrix}$.

Completing the square

halve   square

$$x^2 + 6x = (x + 3)^2 - 9$$

$$\Rightarrow x^2 + 6x + 5 = (x + 3)^2 - 9 + 5 = (x + 3)^2 - 4$$

Roots $x = -\alpha$ and $x = -\beta$ are called the **zeros** of the function $(x + \alpha)(x + \beta)$ and the **roots** of the equation $(x + \alpha)(x + \beta) = 0$.

■ (a) Express $x^2 + 2x - 15$

    (i)   in completed square form;

    (ii)   in factorised form.

(b) Sketch the graph of $y = x^2 + 2x - 15$

● (a) (i)

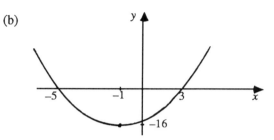

$$x + 2x \quad = (x + 1)^2 - 1$$
$$\Rightarrow x^2 + 2x - 15 = (x + 1)^2 - 16$$

    (ii) $x^2 + 2x - 15 = (x + 5)(x - 3)$

(b)

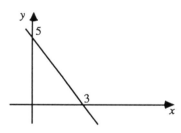

■ $x = -1$ and $x = 5$ are roots of the equation $x^2 + bx + c = 0$.

(a) Find b and c.

(b) Express $x^2 + bx + c = 0$ in completed square form.

(c) Find the equation of the line of symmetry of the graph of $y = x^2 + bx + c$.

● (a) $x^2 + bx + c = (x + 1)(x - 5)$
$$= x^2 - 4x - 5$$

    $b = -4, c = -5$

(b) $x^2 - 4x - 5 = (x - 2)^2 - 9$

(c) The vertex is $(2, -9)$. The line of symmetry is $x = 2$.

---

1 Find the equation of the straight line joining $(4, 3)$ to $(9, 1)$.

2 Sketch the graph of:

(a) $\dfrac{x}{5} + \dfrac{y}{7} = 1$ (b) $y = 10\,000 - x^2$

3 Suggest suitable equations for each of the following graphs:

(a)

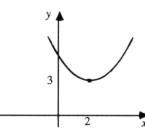

(b)

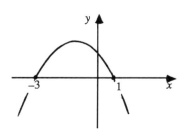

(c)

4 (a) Express $x^2 + 6x - 7$

    (i)   in completed square form;

    (ii)   in factorised form.

(b) Sketch the graph of $y = x^2 + 6x - 7$.

(c) Write down the equation of the line of symmetry of the graph.

5 (a) Express $x^2 - 8x + 25$ in completed square form and sketch the graph of $y = x^2 - 8x + 25$.

(b) Explain why it is not possible to factorise $x^2 - 8x + 25$.

6 The graph $y = a(x - b)^2 + c$ passes through the point $(8, 19)$ and has $(5, 10)$ as its minimum point. Find a, b and c.

7 Factorise:

(a) $x^2 + 8x + 7$ (b) $x^2 - 10x + 21$

(c) $(x + 3)^2 - 4$ (d) $x^2 - 14x + 49$

(e) $p^2 - 16$ (f) $t^2 - 2t - 15$

8 Find the roots of the equations:

(a) $x^2 - 8x = 0$ (b) $x^2 + 6x - 7$

(c) $x^2 + 4x + 4 = 0$ (d) $(x - 5)^2 - 16$

(e) $2x^2 + 8x + 6 = 0$ (f) $x^2 + x - 12$

Sequences

A sequence may be defined in one of two ways:

- **inductively**, using a starting value (or values) and a recurrence relation, for example, $u_{i+1} = 3u_i + 5$, $u_1 = 7$

- using a **general term**, for example, $u_i = 3i^3$

A **periodic** sequence is one whose terms repeat at regular intervals. Other sequences may either converge to a limit or **diverge.**

Sigma notation

Sigma (Σ) notation may be used to express the sum of a series. For example,

$$\sum_{i=2}^{6} (3i + 4) \text{ gives the sum } 10 + 13 + 16 + 19 + 22 = 80$$

A useful device for expressing the general term of a sequence with alternating signs is to use $(-1)^i$ or $(-1)^{i+1}$. For example,

$$\sum_{1}^{4} (-1)^{i+1} 2^i = 2 - 4 + 8 - 16 = -10$$

Arithmetic series

$$a + (a + d) + (a + 2d) + \ldots + (a + (n - 1)d)$$

where a = 1st term
d = common difference
n = number of terms

e.g. $2 + 6 + 10 + 14$ $a = 2, d = 4, n = 4$

$5 + 4\frac{1}{2} + 4 + 3\frac{1}{2} + 3$ $a = 5, d = -\frac{1}{2}, n = 5$

The sum of an arithmetic series is given by:

number of terms x average of first and last terms

Geometric series

$$a + ar + ar^2 + \ldots + ar^{n-1}$$

where a = 1st term
r = common ratio
n = number of terms

e.g. $2 + 6 + 18 + 54$ $a = 2, r = 3, n = 4$

$1 + \frac{1}{2} + \frac{1}{4} + \frac{1}{8} + \frac{1}{16}$ $a = 1, r = \frac{1}{2}, n = 5$

$$\sum_{1}^{n} ar^{i-1} = a + ar + ar^2 + \ldots + ar^{n-1}$$

$$= a\frac{(1-r^n)}{1-r} \text{ or } a\frac{(r^n - 1)}{r - 1}$$

For $|r| < 1$, the sum of an infinite G.P. is

$$a + ar + ar^2 + \ldots = \frac{a}{1-r}$$

■ Sum the series:

   (a) $1 + 3 + 5 + \ldots + 99$

   (b) $4 + 9 + 14 + \ldots$ as far as the 40th term

● (a) $a = 1, d = 2$

      $a + (n - 1)d = 99 \Rightarrow n = 50$

      $\text{Sum} = 50 \times \dfrac{1 + 99}{2} = 2500$

   (b) $a = 4, d = 5$

      $a + 39d = 199$

      $\text{Sum} = 40 \times \dfrac{4 + 199}{2} = 4060$

■ (a) Find the sum to infinity of

      $1 + \dfrac{2}{3} + \dfrac{4}{9} + \dfrac{8}{27} + \ldots$

   (b) How many terms of the series must be taken before its sum to n terms differs from its sum to infinity by less than 0.1?

● (a) The sum to infinity is 3.

   (b) The sum to n terms is

      $\dfrac{1 - (\frac{2}{3})^n}{1 - \frac{2}{3}} = 3\left(1 - (\tfrac{2}{3})^n\right)$

      The difference, $3(\tfrac{2}{3})^n$, is less than 0.1 when $n = 9$.

---

1 Calculate the sum of the series:

   (a) $12 + 16 + 20 + \ldots$ to 24 terms

   (b) $16 + 11 + 6 + 1 + \ldots$ to 35 terms

   (c) $1 + 2 + 4 + 8 + \ldots$ to 21 terms

   (d) $1 + 1.04 + (1.04)^2 + (1.04)^3 + \ldots + (1.04)^{17}$

2 Rewrite using Σ notation:

   (a) $\dfrac{1}{2} + \dfrac{1}{5} + \dfrac{1}{8} + \dfrac{1}{11} + \ldots + \dfrac{1}{56}$

   (b) $\dfrac{1}{2} + \dfrac{2}{3} + \dfrac{3}{4} + \dfrac{4}{5} + \ldots + \dfrac{87}{88}$

   (c) $2^3 + 4^3 + 6^3 + \ldots + 100^3$

3 For each of the following sequences, write down the first five terms and the value of the 30th term:

   (a) $u_1 = -1,\qquad u_{k+1} = u_k - 2$

   (b) $u_1 = 4,\qquad u_{k+1} = 2u_k$

   (c) $u_1 = 5,\qquad u_{k+1} = \dfrac{1}{u_k}$

4 Let u_n be the number of ways of arranging n objects in a line. For example:

$u_3 = 6$

   (a) Find u_2 and u_4.

   (b) Explain why $u_{n+1} = (n + 1)u_n$.

   (c) Calculate u_{10}.

5 Calculate the sum $\displaystyle\sum_{1}^{\infty} \dfrac{1}{2^{i-1}}$

6 Find the sum of the first $2n$ terms of the G.P. $5, 25, 125, \ldots$

7 Find the sum of the integers between 220 and 541 which are divisible by 9.

8 (a) $\displaystyle\sum_{i=1}^{n} (3i + k) = 8 + 11 + 14 + 17 + \ldots + 155$

      Find k and n.

   (b) Find the sum of the series in (a).

9 Find the sum of the infinite series

   (a) $7 + \dfrac{7}{2} + \dfrac{7}{4} + \dfrac{7}{8} + \ldots$

   (b) $3 - 2 + \dfrac{4}{3} - \dfrac{8}{9} + \ldots$

10 A pattern of dots is as follows:

   (a) How many dots are there on the perimeter of the nth diamond?

   (b) If there are d_n dots in the nth diamond, then explain why

      $$d_{n+1} = d_n + 4n$$

   (c) Explain why

      $$d_n = 1 + 4(1 + 2 + \ldots + (n - 1))$$

      Hence find d_n.

Notation   The definition of a function has the form:

$$f(x) = \sqrt{(x-5)}, \ x \geq 5$$

The **rule** tells you how the function is to be calculated and the **domain** tells you the set of values to which the rule may be applied.

In describing the domain, the following notation for certain important sets of numbers is useful.

Z   the integers ... $-3, -2, -1, 0, 1, 2, ...$
N   the natural numbers $1, 2, 3, 4, ...$
Q   the rationals or fractions
R   the real numbers.

Translations   The graph of $f(x+a)+b$ is a translation of the graph of $f(x)$ through $\begin{bmatrix} -a \\ b \end{bmatrix}$.

For example,

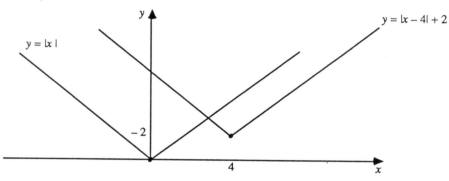

Dominance   A graph of a polynomial function of the form

$$a + bx + ... + x^n$$

looks like the graph of $y = a + bx$ for very small x and like the graph of $y = x^n$ for very large x.

Thus the graph of $y = x^3 - 2x$ has the form:

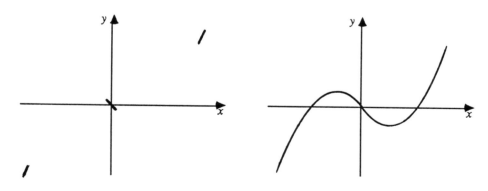

Finding the roots can help establish the precise shape of the graph.

■ Sketch $f(x) = x^2(x - 1)(x + 4)$.

● For large x, $f(x) \approx x^4$
For small x, $f(x) \approx -4x^2$

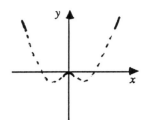

The zeros are at $x = 0$, 1 and -4.

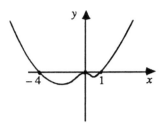

■ (a) Sketch the graph of $y = \sqrt{(x + 4)} - 2$.

(b) What is the largest possible domain for
$f(x) = \sqrt{(x + 4)} - 2$?

● (a) The graph is obtained by translating the graph of
$y = \sqrt{x}$ by $\begin{bmatrix} -4 \\ -2 \end{bmatrix}$.

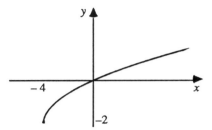

(b) The largest possible domain is $x \geq -4$.

1 (a) For the function $f(x) = x^3 + x^2 + \dfrac{1}{\sqrt{x}}$, find:

(i) $f(4)$ (ii) $f(1)$ (iii) $f(0.5)$

(b) What is the largest possible domain for $f(x)$?

2 Find the zeros of the function

$$f(x) = x(x - 3)^2$$

and hence sketch the curve.

3 Sketch the curve

$$f(x) = (x - 9)(x + 1)(2x + 3)$$

describing and explaining the main features.

4 Suggest suitable equations for the following graphs:

(a)

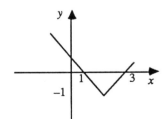

(b)

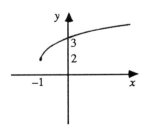

5 (a) For the function $f(x) = x^3$, find the function
$g(x) = f(x - 2) + 9$.

(b) State how the two graphs of $f(x)$ and $g(x)$ are related.

(c) Sketch the graph of $g(x)$.

6 (a) What term is suggested by the shape of the graph given below when x is large?

(b) What term is suggested by the shape of the graph when x is small?

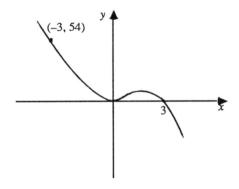

(c) Use the zero to obtain a possible equation for the graph. Ensure that the equation gives $y = 54$ when $x = -3$.

(d) Suggest a possible equation for a similarly shaped graph passing through $(-3, 108)$, $(0, 0)$ and $(3, 0)$.

Algebra

Letters can be used to:

- represent unknown quantities;
- state general results;
- stand for independent and dependent variables.

Identities

An identity is true for any value of its variables. Some useful identities are:

$$a^2 + 2ab + b^2 = (a + b)^2$$
$$a^2 - 2ab + b^2 = (a - b)^2$$
$$a^2 - b^2 = (a - b)(a + b)$$

Inequalities

These can be treated in the same way as equations except that whenever the inequality is multiplied or divided throughout by a negative quantity, the sense of the inequality is changed.

■ Solve $3t + 1 > 5t - 8$.

●
$$3t + 1 > 5t - 8$$
$$\Rightarrow \quad -2t > -9$$
$$\Rightarrow \quad t < \frac{9}{2}$$

The solution of an inequality can be deduced from the solution of the corresponding equation. A sketch can help you make the correct deduction.

For example,

$$3t + 1 = 5t - 8$$
$$\Rightarrow \quad 9 = 2t$$
$$\Rightarrow \quad t = \frac{9}{2}$$

So $3t + 1 > 5t - 8 \Rightarrow t < \frac{9}{2}$

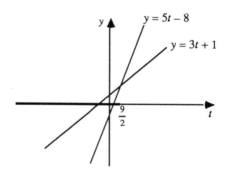

Quadratic equations

$ax^2 + bx + c = 0$ may be solved by:

(a) factorising,

(b) completing the square,

(c) the formula $x = \dfrac{-b \pm \sqrt{(b^2 - 4ac)}}{2a}$, (for $a \neq 0$).

If $b^2 > 4ac$ the equation will have two roots.

If $b^2 = 4ac$ the equation will have a single (repeated) root.

If $b^2 < 4ac$ the equation will have no roots.

Factor theorem

A **polynomial** is a function involving whole number powers of a variable. The highest power present is called the **degree** of the polynomial. For example, a cubic polynomial is a polynomial of degree 3, such as

$$P(x) = x^3 - 5x + 4$$

The zeros of a polynomial, $P(x)$, are values of x such that $P(x) = 0$. For the polynomial given above, 1 is a zero because $P(1) = 1^3 - 5 \times 1 + 4 = 0$.

If $x = a$ is a solution of the polynomial equation $P(x) = 0$, then $x - a$ is a factor of $P(x)$. For the polynomial given above, $P(1) = 0$ and so $x - 1$ is a factor. In fact:

$$P(x) = (x - 1)(x^2 + x - 4)$$

The **factor theorem** can be used to help factorise a polynomial, providing a simple zero can be found.

■ Solve $x^2 > 4x - 3$.

● $x^2 = 4x - 3 \Rightarrow x^2 - 4x + 3 = 0$
$\Rightarrow x = 1$ or 3

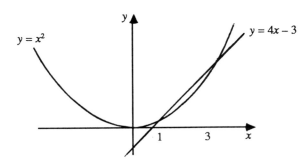

From the graph, the inequality is true when $x < 1$ or $x > 3$.

■ Solve $x^3 - 4x^2 - 11x + 30 = 0$.

● By trial, $P(2) = 8 - 16 - 22 + 30 = 0$.
Therefore $x - 2$ is a factor and the remaining factor will be a quadratic.

$$x^3 - 4x^2 - 11x + 30 = (x - 2)(x^2 + ax - 15)$$

Comparing terms in x^2,
$$-4x^2 = -2x^2 + ax^2 \Rightarrow a = -2$$
$$P(x) = (x - 2)(x^2 - 2x - 15)$$
$$= (x - 2)(x - 5)(x + 3)$$
$$x = -3, 2, 5$$

1 Simplify:

  (a) $(x + y)^2 - (x - y)^2$

  (b) $(a - b)(a + b) + b^2$

2 Solve the following equations, giving your answers in exact form.

  (a) $x^2 - 2x - 8 = 0$

  (b) $x^2 - 7x + 3 = 0$

3 The graphs of $y = 12 - 2x$ and $y = x^2 - 7x + 6$ are as shown

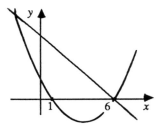

  Use the diagram to help you solve:

  $$x^2 - 7x + 6 < 12 - 2x$$

4 (a) On the same set of axes, sketch the graphs of $y = x^2 + 7x$ and $y = 4x + 4$.

  (b) Use your graph to solve the inequality $x^2 + 7x < 4x + 4$

5 Given that $P(x) = 3x^4 - 6x^3 - 3x^2 + 6x$:

  (a) find $P(-2)$, $P(-1)$, $P(1)$ and $P(2)$;

  (b) hence, completely factorise the polynomial.

6 $P(x) = x^3 - 5x^2 - 4x + 20$
  Given that $P(5) = 0$, solve $P(x) \geq 0$.

7 For the polynomial:

  $$P(x) = x^3 + 6x^2 - 2x - 5$$

  (a) list all the possible factors of $P(x)$;

  (b) factorise $P(x)$ as far as possible;

  (c) solve the equation $P(x) = 0$.

8 If $P(x) = x^3 + x^2 - 14x - 24$:

  (a) factorise $P(x)$ completely;

  (b) find all the solutions of $P(x) = 0$;

  (c) sketch the graph of $P(x)$;

  (d) solve the inequality $x^3 + x^2 - 14x - 24 \geq 0$.

9 Cola has been spilt on the page of a mathematics book and has obliterated some of the text.

  The following extract shows the method of solving a quadratic equation by completing the square:

  $$x^2 + 6x - 5 = 0$$
  $$(x \quad)^2 \, . \, . = 0$$
  $$. \quad . \quad . \quad . \quad . \quad .$$
  $$. \quad . \quad . \quad . \quad . \quad .$$
  $$x = -3 \pm \sqrt{14}$$

  Suggest how the text should be restored by copying and completing the mathematical argument.

Locating the roots

In order to solve an equation numerically you can use either **binary/decimal search** or an **iterative formula**. For both methods you must first locate the roots. This can be done in one of a number of ways.

To solve $f(x) = 0$, you can sketch the graph of $y = f(x)$ and note where it crosses the x-axis.

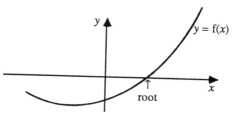

Alternatively, to solve $f(x) = 0$ you can use the property that if the sign of $f(a)$ differs from that of $f(b)$, then (providing the graph of $f(x)$ is continuous) there is a root in the interval $[a, b]$.

To solve $f(x) = g(x)$ you can sketch the graphs of $y = f(x)$ and $y = g(x)$ and note where they intersect.

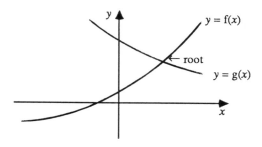

Zoom

One of the simplest methods of finding roots is to use the zoom and trace facilities on a graphic calculator.

Decimal and binary search

If, for example, a root is in the interval $[2, 3]$ the method of **decimal search** requires you to evaluate $f(x)$ at $x = 2.1, 2.2, 2.3, \ldots$ until the sign changes. This can be repeated for smaller intervals. If the root lies in $[2.4, 2.5]$ you would consider $x = 2.41, 2.42, 2.43, \ldots$ until the required accuracy is achieved.

The method of **binary search** just involves halving the interval each time. For example, having located a root in $[2, 3]$, you would evaluate $f(2.5)$ and then search either in $[2, 2.5]$ or $[2.5, 3]$.

Iterative formula

If the equation $f(x) = 0$ is rearranged as $x = g(x)$, the iterative formula $x_{i+1} = g(x_i)$ may converge to a solution (though in many cases it may not).

■ Show that the equation $x^2 = 2$ can be rearranged as

(a) $\quad x = \dfrac{2}{x}$

(b) $\quad x = \dfrac{2}{3}\left(x + \dfrac{1}{x}\right)$

In each case, use an iteration with $x_1 = 1.5$ to attempt to find a value for $\sqrt{2}$ to 2 decimal places.

● (a) $\quad x^2 = 2 \Rightarrow x = \dfrac{2}{x}$ (Divide by x)

Then $x_1 = 1.5, x_2 = 1.3, x_3 = 1.5, x_4 = 1.3, \ldots$

The sequence oscillates and does not converge to a solution.

(b) $\quad x^2 = 2 \Rightarrow \dfrac{1}{3}x = \dfrac{2}{3x} \Rightarrow x = \dfrac{2}{3}\left(x + \dfrac{1}{x}\right)$ $\quad\left(+ 2x^2\right)$

Then $x_1 = 1.5, x_2 = 1.444\,444\,44, x_3 = 1.424\,501\,425, x_4 = 1.417\,667\,617,$

$\qquad x_5 = 1.415\,367\,719, x_6 = 1.414\,59.$

$\sqrt{2} \approx 1.41$

■ Solve the equation $xe^{0.5x} = 2$ (to 2 d.p.).

● By sketching the graphs of $y = xe^{0.5x}$ and $y = 2$, the root can be seen to lie between 1 and 2.

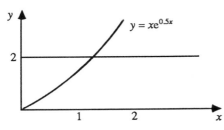

Rearranging gives $x = 2e^{-0.5x}$, so $x_{i+1} = 2e^{-0.5x_i}$

Let $x_0 = 1$
$x_1 = 1.213$, $x_2 = 1.090$, ..., $x_9 = 1.1338$, $x_{10} = 1.1345$
Since x_9 and x_{10} agree to 2 decimal places, $x \approx 1.13$.

■ For $f(x) = x^2 - 2 - \dfrac{1}{x+3}$:

(a) prove that $x = 1$ and $x = 2$ are bounds for a root of $f(x) = 0$;

(b) use decimal or binary search to calculate, to 1 decimal place, the value of the solution between $x = 1$ and $x = 2$.

● (a) $f(1) = -1.25$, $f(2) = 1.8$ and so there is a root in $[1, 2]$.

(b) ..., $f(1.4) = -0.267$, $f(1.5) = 0.278$, ...

$f(1.45) = -0.122$ and so the root is in $[-1.45, 1.5]$

The root is 1.5 (to 1 d.p.).

---

1 For each of the following, sketch appropriate graphs and find bounds for all possible solutions.

(a) $x^2 - 5 = 0$

(b) $x^3 - x^2 - 3x + 1 = 0$

(c) $x^3 = 3x + \dfrac{1}{x}$

2 Show how each rearrangement may be found from the given equation.

(a) $x = \dfrac{1}{6}(8 - x^3)$ from $x^3 + 6x - 8 = 0$

(b) $x = \dfrac{2x - 5}{x^2}$ from $x^3 - 2x + 5 = 0$

(c) $x = (3 - 2x)^2$ from $2x + \sqrt{x} - 3 = 0$

(d) $x = \dfrac{1}{x^2} - \dfrac{2}{x}$ from $x^2 + 2 - \dfrac{1}{x} = 0$

3 (a) Use a graph plotter to estimate the roots of $x^3 - 8x - 7 = 0$ to 1 decimal place.

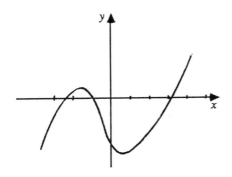

(b) Factorise $x^3 - 8x - 7$.

(c) Hence find precise values for the three roots.

4 In the graph shown below, $f(1) = -3$, $f(2) = 5$.

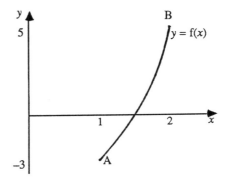

By joining the points A and B by a straight line, obtain a first approximation to the root of $f(x) = 0$.

5 For the equation $x^3 = 7$:

(a) sketch the curve $y = x^3 - 7$ and hence state an interval that contains the root of $x^3 = 7$;

(b) show that the equation can be rearranged to $x = \sqrt{\sqrt{(7x)}}$.

(c) By letting $x_1 = 2$ and using an iterative formula, obtain the solution for $x^3 = 7$ correct to 3 decimal places.

6 (a) By means of a sketch graph determine the number of roots of the equation $x = 1.5 \sin x$.

(b) Use an iterative formula to find these roots correct to 3 decimal places.

INTRODUCTORY CALCULUS

Rate of change The rate of change of one variable with respect to another is the change in the first variable per unit increase in the second.

A negative rate of change means the first variable decreases as the second increases. For a straight line graph of y against x, the rate of change of y with respect to x, $\frac{dy}{dx}$, is the gradient of the straight line.

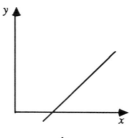 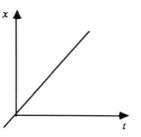 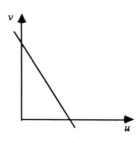

gradient, $\frac{dy}{dx}$, is positive   gradient, $\frac{dx}{dt}$, is positive   gradient, $\frac{dv}{du}$, is negative

Gradients If a curve appears to be a straight line when you zoom in at a point, then it is said to be 'locally straight' at that point. The gradient of the tangent to the curve at a point is equal to the gradient of the locally straight line at that point.

The gradient of the tangent to a curve, $\frac{dy}{dx}$, represents the $\dfrac{\text{difference between the } y \text{ coordinates}}{\text{difference between the } x \text{ coordinates}}$ along the tangent at P.

If you plot the gradient at each point on a curve against x, you obtain the **gradient graph**.

Points of a graph where the curve has zero gradient are called stationary points e.g. B, D, F, G, H.

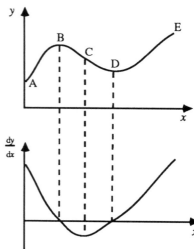

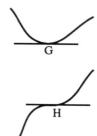

B, D, F and G are called **turning points**.
B and F are **local maxima**.
D and G are **local minima**.

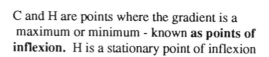

C and H are points where the gradient is a maximum or minimum - known **as points of inflexion.** H is a stationary point of inflexion

Differentiation Differentiation is the process of finding the **derivative** of a function, called its **derived function.**

The derivative of y with respect to x is written as $\frac{dy}{dx}$ or $\frac{d}{dx}(y)$.

$\frac{d}{dx}(f(x))$ is usually written as $f'(x)$.

If $y = a + bx + cx^2 + dx^3$, then $\frac{dy}{dx} = b + 2cx + 3dx^2$.

■ If $y = 7x^3$, find $\dfrac{dy}{dx}$.

● $\dfrac{dy}{dx} = 7\,(3x^2) = 21x^2$

■ Find the equation of the tangent to $y = 3 + 2x + x^2$ at the point $(1, 6)$.

● $\dfrac{dy}{dx} = 2 + 2x$

At the point $(1, 6)$, $\dfrac{dy}{dx} = 2 + 2 \times 1 = 4$.

The tangent has gradient 4 and passes through $(1, 6)$ so the equation is

$$\dfrac{y-6}{x-1} = 4$$

$$\Rightarrow y = 4x + 2$$

■ Find the range of values of x for which $y = x^2 - 6x + 3$ is decreasing.

● $\dfrac{dy}{dx} = 2x - 6$

$2x - 6 < 0 \Rightarrow x < 3$.

The function is decreasing for $x < 3$.

■ If $y = 6x^2 - 7x + 8$, find $\dfrac{dy}{dx}$.

● $\dfrac{dy}{dx} = 6\,(2x) - 7 = 12x - 7$

■ Use a numerical method to estimate the derivative of $y = 3^x$ at the point $(2, 9)$.

●
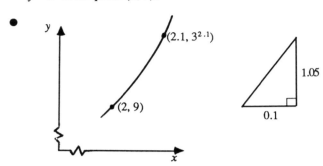

$3^{2.1} - 9 \approx 1.05$.

The derivative is approximately $\dfrac{1.05}{0.1} = 10.5$.

[Using smaller intervals gives answers closer to the actual derivative of 9.89 (2 d.p.)]

---

1 (a) Use the sketch to write down $\dfrac{ds}{dt}$.

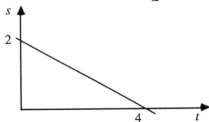

   (b) Hence write down the equation of the line.

2 A linear graph has $\dfrac{dp}{dr} = 2$ and passes through the point $(1, 3)$. Find its equation.

3 Firm A charges a basic call out fee of £15 plus £6 per hour for each plumber sent out for emergency repairs.

   (a) If one plumber is called out for t hours then write down an expression for the charge £u in terms of t.

   A rival firm, B, charges a basic fee of £10 plus £8 per hour. Write down an expression for the charge £v if one plumber is called out for t hours.

   (b) Comment on which firm is cheaper.

   (c) Write down $\dfrac{du}{dt}$ and $\dfrac{dv}{dt}$ and explain their significance.

   (d) In an emergency a factory is able to call out 4 plumbers from A and 3 from B. Write down the total cost £C in terms of u and v.

   Deduce the value of $\dfrac{dC}{dt}$ and explain its meaning.

4 For the curve given below

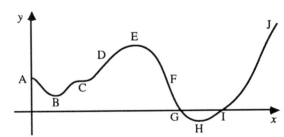

   (a) sketch the gradient curve;

   (b) identify (i)   the stationary points;
                  (ii) the turning points;
                  (iii) the local minima;
                  (iv) the overall maximum;
                  (v)   the zeros.

5 Find $\dfrac{dy}{dx}$ when (a) $y = 5x^3 + x$ (b) $y = 8 - x + 3x^2$

6 Find the equation of the tangent to the curve $y = x^3 - 3x + 5$ at the point $(2, 7)$.

Stationary points   For the graph of a function, the stationary points can be found by equating the derived function to zero.

You can distinguish between local maxima, local minima and stationary points of inflection by considering the signs of the derivative on either side of these values.

Local minimum   **Local maximum**   **Stationary points of inflexion**

$\frac{dy}{dx} < 0 \quad \frac{dy}{dx} > 0$ $\frac{dy}{dx} > 0 \quad \frac{dy}{dx} < 0$ $\frac{dy}{dx} < 0 \quad \frac{dy}{dx} < 0 \quad \frac{dy}{dx} > 0 \quad \frac{dy}{dx} > 0$

Graph sketching   To sketch the graph of a quadratic or cubic function

(a) The sign of the highest power of x indicates the general shape,

e.g. x^2  $-x^2$

x^3  $-x^3$

(b) Find the turning points.

(c) Find the point at which the graph crosses the y-axis, by putting $x = 0$ into the function.

(d) Do not attempt to fully label axes. Only 'critical' values need be marked.

(e) If the function can be factorised, mark the points at which the graph crosses the x-axis.

Optimisation problems   To find the maximum/minimum values of a quantity

(a) Express the quantity in terms of a single variable.

(b) Consider the turning points of the graph of this relationship. If these cannot be found algebraically, use a graph plotter and 'estimate' the values.

■ Find the maximum value of xy given that $x + y = 10$.

● $xy = x(10 - x) = 10x - x^2$.

This has turning points when
$$10 - 2x = 0$$
i.e. $x = 5$ and $y = 10 - x = 5$.

xy is then 25 and this is clearly a maximum.

■ Sketch the graph of $y = x^3 - 3x^2 - 9x + 7$.

● $\dfrac{dy}{dx} = 3x^2 - 6x - 9 = 3(x-3)(x+1)$

$\dfrac{dy}{dx} = 0$ when $x = 3$ or -1

When $x = 3$, $y = -20$. When $x = -1$, $y = 12$.

The graph is cubic with shape as shown.

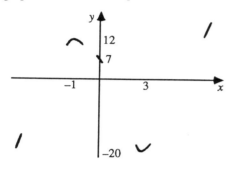

Thus $(3, -20)$ is a minimum and $(-1, 12)$ is a maximum.

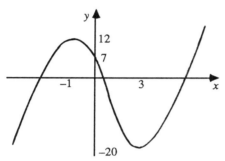

■ An industrialist acquires a piece of land OABC with the dimensions shown. He wants to build a unit with a rectangular base with the largest area. Will this be the rectangle ODBC or a rectangle like OEFG?

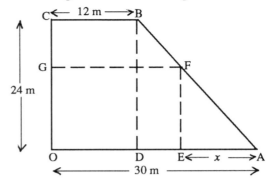

● Let EA $= x$. Then, by similar triangles:

$$\dfrac{FE}{EA} = \dfrac{FE}{x} = \dfrac{BD}{DA} = \dfrac{24}{18} \Rightarrow FE = \dfrac{4}{3}x$$

The area of rectangle OEFG, A, is

$\dfrac{4}{3}x(30-x) = \dfrac{4}{3}(30x - x^2)$

$\Rightarrow \dfrac{dA}{dx} = \dfrac{4}{3}(30 - 2x) = 0$ when $x = 15$.

x	$\rightarrow$	15	$\rightarrow$
$\dfrac{dA}{dx}$	+	0	−

$x = 15$ gives max $A = 300$ m²
Area of rectangle ODBC $= 12 \times 24 = 288$ m² $<$ max A

---

1 $f(x)$

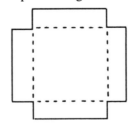

State the signs of $f(x)$ and $f'(x)$ inside each of the intervals A, B, C, D, E in the diagram.

2 For the graph of $y = x^3 - \dfrac{3}{2}x^2 - 6x$, which of the following points are turning points?

A $(0, 0)$ B $(-1, \dfrac{7}{2})$ C $(1, -\dfrac{13}{2})$ D $(2, -10)$

3 The velocity v m s⁻¹ of a body moving in a straight line is given by $v = 6t - 3t^2$ where t is the time in seconds. Find the maximum velocity.

4 Find and describe the stationary points of:

(a) $y = x^4 - 1$ (b) $y = x^2e^x$

5 Find the local maximum and minimum points of the function
$$y = 3x^4 + 8x^3 - 18x^2 + 12$$

6 A cylinder is such that the sum of its height and the circumference of its base is 6 m. Find the greatest possible value of the volume of the cylinder.

7 From a 12 cm by 12 cm square piece of cardboard, a square of side x cm is cut from each corner, as shown in the diagram. The remaining cardboard is then folded to form an open rectangular box of depth x cm.

(a) Show that the volume V of the box is given by
$V = 4x(6 - x)^2$.

(b) If the volume of the box is to be as large as possible, find x and hence state the dimensions of the box.

Definite integrals

$\int_a^b f(x)\,dx$ is the area under the curve $y = f(x)$ between $x = a$ and $x = b$.

$\int_a^b f(x)\,dx$ treats areas **beneath** the x-axis as negative

i.e. $\int_a^b y\,dx = A - B$.

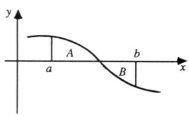

Mid-ordinate rule

The mid-ordinate rule uses a series of rectangles to estimate the area under a graph. The height of each rectangle is determined by the height of the curve at the mid-point of the interval.

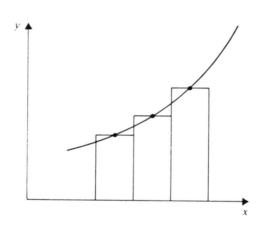

■ Evaluate $\int_3^5 \sqrt{(x^2 + 1)}\,dx$ using the mid-ordinate rule with 4 strips.

● Each strip has width $\frac{1}{2}$ unit.

The mid-points and heights are:

x	3.25	3.75	4.25	4.75
height	3.40	3.88	4.37	4.85

Area $\approx \frac{1}{2} \times (3.40 + 3.88 + 4.37 + 4.85)$

$\int_3^5 \sqrt{(x^2 + 1)} \approx 8.25$

The accuracy of the solution can be improved by taking a larger number of strips.

Trapezium rule

The trapezium rule uses a series of trapezia to estimate the area under a graph.

The area of a trapezium is $\frac{1}{2}(y_1 + y_2)h$.

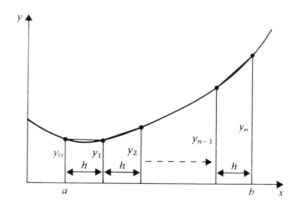

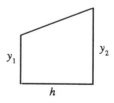

The trapezium rule can be summarised as

$$\int_a^b y\,dx \approx \frac{1}{2}h(y_0 + y_1) + \frac{1}{2}h(y_1 + y_2) + \ldots + \frac{1}{2}h(y_{n-1} + y_n)$$

■ Find $\int_0^1 \sqrt{(1 + x^2)}\,dx$ using the mid-ordinate rule with 5 strips.

● $f(x) = \sqrt{(1 + x^2)}$

x	0.1	0.3	0.5	0.7	0.9
$f(x)$	1.005	1.044	1.118	1.221	1.345

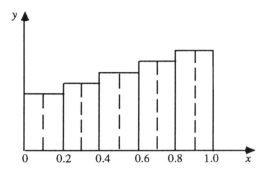

The total area is

0.2 x 1.005 + 0.2 x 1.044 + 0.2 x 1.118 + 0.2 x 1.221 + 0.2 x 1.345
= 0.2 x 5.733 = 1.1466 ≈ 1.15

■ Find $\int_2^4 \sqrt{(x^3 - 1)}\,dx$ using the trapezium rule with 4 strips.

● If $f(x) = \sqrt{(x^3 - 1)}$ you can tabulate

x	2	2.5	3.0	3.5	4.0
$f(x)$	2.646	3.824	5.099	6.471	7.937

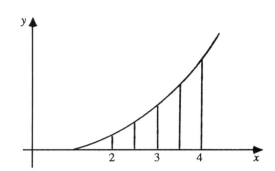

The total area is

$\frac{1}{2}(2.646 + 3.824) \times 0.5 + \frac{1}{2}(3.824 + 5.099) \times 0.5 +$

$\frac{1}{2}(5.099 + 6.471) \times 0.5 + \frac{1}{2}(6.471 + 7.937) \times 0.5$

≈ 10.34

---

1 The circumference, c m, of a tree trunk is measured at various heights, h m, as follows:

h m	0	1	2	3	4	5	6
c m	2.53	2.21	1.86	1.39	1.04	0.67	0.42

(a) Assuming that the trunk is roughly circular, estimate the cross-sectional area at each point.

(b) Estimate the total volume of timber in the trunk

    (i) using the mid-ordinate rule with 3 strips;

    (ii) using the trapezium rule with 6 strips.

2 Evaluate $\int_3^5 \sqrt{(x^2 + 1)}\,dx$ using

(a) the mid-ordinate rule with 5 strips;

(b) the trapezium rule with 5 strips.

3 (a) Estimate the area under the graph of $f(x) = \sqrt{(1 + \frac{1}{x})}$ between $x = 1$ and $x = 3$.

(b) Is your estimate an over- or under-estimate of the true value? Justify your answer.

4 By sketching the graph of $f(x) = \cos x$, state the value of $\int_0^\pi \cos x\,dx$.

5 Find the smallest value of a for which $\int_0^a \sin 3x\,dx = 0$.

6

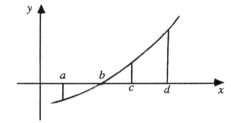

It is given that $\int_a^b f(x) = A$, $\int_b^c f(x)\,dx = B$, $\int_c^d f(x)\,dx = C$

State the value of:

(a) $\int_b^d f(x)\,dx$ (b) $\int_a^c f(x)\,dx$ (c) $\int_a^d f(x)\,dx$

in terms of A, B and C.

7

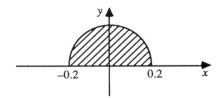

A steel bar, 4 m long is in the shape of the curve $y = \sqrt{(0.04 - x^2)}$ which lies above the x-axis. If 1 m³ of steel weighs 8 tonnes, estimate the weight of the bar.

Area functions

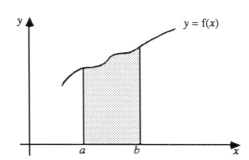

$$\int_a^b f(x)\, dx = A(b) - A(a)$$

$A(x)$ is known as an area function or as an **integral function**.

A polynomial function $f(x) = a + bx + cx^2 + dx^3$
has integral function $\quad A(x) = ax + \frac{1}{2}bx^2 + \frac{1}{3}cx^3 + \frac{1}{4}dx^4$

Definite integrals

The area under the graph of $y = f(x)$ between $x = a$ and $x = b$ is evaluated using the definite integral $\int_a^b f(x)\, dx$.

■ Evaluate $\int_2^4 (5x^2 + 3)\, dx$

● $\int_2^4 (5x^2 + 3)\, dx = \left[\frac{5}{3}x^3 + 3x \right]_2^4$

$\qquad = \left(\frac{5 \times 4^3}{3} + 3 \times 4 \right) - \left(\frac{5 \times 2^3}{3} + 3 \times 2 \right)$

$\qquad = 99\frac{1}{3}$

The fundamental theorem of calculus

$$\int_a^b f'(x)\, dx = f(b) - f(a)$$

This result demonstrates that integration and differentiation are inverse processes.

■ If $\int_a^b g(x)\, dx = \left[x^3 - 5x \right]_a^b$, then find $g(x)$.

● $g(x) = \frac{d}{dx}(x^3 - 5x) = 3x^2 - 5$

Indefinite integrals

$\int f(x)\, dx$ is known as the indefinite integral of $f(x)$ and includes a constant of integration, c.

$$\int x^n\, dx = \frac{x^{n+1}}{n+1} + c, \text{ for } n \neq -1$$

For any constant a, and any functions f and g,

$$\int a\, f(x)\, dx = a \int f(x)\, dx \qquad \int (f(x) + g(x))\, dx = \int f(x)dx + \int g(x)\, dx$$

Constants of integration can be determined if you have additional information about the integral function.

■ Find y as a function of x given that $y = 5$ when $x = 1$ and that $\frac{dy}{dx} = (3x - 1)(x + 5)$

● $\qquad y = \int (3x - 1)(x + 5)\, dx = \int (3x^2 + 14x - 5)\, dx$

$\qquad y = x^3 + 7x^2 - 5x + c$

Then $5 = 1 + 7 - 5 + c \Rightarrow c = 2$

$\qquad y = x^3 + 7x^2 - 5x + 2$

■ Find the shaded area.

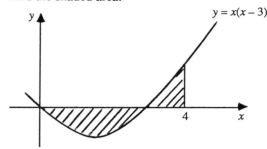

● $y = x(x - 3)$ crosses the x-axis at $x = 0$ and $x = 3$.

To find the area you need to calculate two integrals.

$$\int_0^3 x(x - 3)\,dx = \int_0^3 (x^2 - 3x)\,dx$$
$$= \left[\frac{1}{3}x^3 - \frac{3x^2}{2}\right]_0^3 = \left(9 - \frac{27}{2}\right) - 0 = -\frac{9}{2}$$

$$\int_3^4 x(x - 3)\,dx = \int_3^4 (x^2 - 3x)\,dx$$
$$= \left[\frac{1}{3}x^3 - \frac{3x^2}{2}\right]_3^4 = \frac{11}{6}$$

Hence the total area is $\frac{9}{2} + \frac{11}{6} = \frac{19}{3}$.

■ Find the constant a, given that the shaded region has area 4 square units.

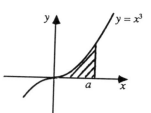

● $\int_0^a x^3\,dx = \left[\frac{1}{4}x^4\right]_0^a = \frac{1}{4}a^4$

$\frac{1}{4}a^4 = 4 \Rightarrow a^4 = 16 \Rightarrow a = 2$

■ Find the value of a such that the two shaded areas are equal.

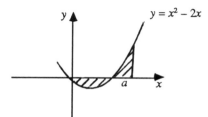

● $0 = \int_0^a (x^2 - 2x)\,dx = \left[\frac{1}{3}x^3 - x^2\right]_0^a$

$a = 3$

1 (a) Find $\int_{-2}^1 (t^3 + 2t^2 - 3)\,dt$.

(b) Explain, using a sketch, why the answer is negative.

2 Calculate the area between the curve and the x- and y-axes in the first quadrant for $y = x^4 - x^3 - 7x^2 + x + 6$.

3 An object starts from rest and its speed, v m s^{-1}, is given by $v = 12t - 3t^2$. Draw a graph to represent this for $0 \le t \le 5$. Calculate the distance travelled in the third second.

4 Find:

(a) $\int x(x^2 - 2)\,dx$ (b) $\int (3x + 2)^2\,dx$

5 A curve passes through the origin and its gradient function is given by $2x - \frac{x^2}{2}$. Find the equation of the curve.

6 Evaluate:

(a) $\int_1^2 x(x - 1)\,dx$ (b) $\int_0^3 (1 - s^2)^2\,ds$

7 If $\int f(t)\,dt = t^2 + 2t + c$, write down f($t$).

8 Find the shaded areas:

(a)

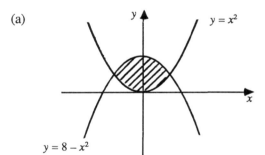

(b)

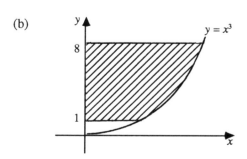

FUNCTIONS

Composition of functions

If $f(x) = 2x - 1$ and $g(x) = x^3$, then:

$$fg(x) = f(x^3) = 2x^3 - 1 \text{ and}$$
$$gf(x) = g(2x - 1) = (2x - 1)^3$$

Domain and range

The set of values for which a function is defined is called its **domain** and the set of values which the function can take is called its **range**. For example, for the function $f(x) = x^2$, the domain is all real numbers and the range is zero and all real positive numbers.

For a function, each member of the domain can lead to only one member of the range.

If only one member of the domain can lead to a given member of the range the function is **one to one**, but if more than one member of the domain leads to the same member of the range the function is **many to one**. (The function $f(x) = x^2$ is many to one.)

Inverse functions and formulas

f^{-1} is the inverse function to f. If $f(x) = 3x$, $f^{-1}(x) = \frac{1}{3}x$. The graph of $y = f^{-1}(x)$ is the graph of $y = f(x)$ after reflection in $y = x$.

Rearranging formulas is a process similar to finding inverse functions.

Flow diagrams can be used if the new subject appears once only on the right-hand side.

■ Make t the subject of $v = u + at$.

● $t \rightarrow \boxed{\times a} \rightarrow \boxed{+ u} \rightarrow v$

$\dfrac{v - u}{a} \leftarrow \boxed{\div a} \leftarrow \boxed{- u} \leftarrow v$

$$t = \frac{v - u}{a}$$

Parameters

The formula $v = u + at$ enables you to work out the velocity (v) of an object at time t. The values of u (initial velocity) and a (acceleration) are fixed for a particular movement, so **u** and **a** are called **parameters**.

Transformations of graphs

The graph of $y = f(x + p)$ is the image of the graph of $y = f(x)$ after translation through the vector $\begin{bmatrix} -p \\ 0 \end{bmatrix}$.

The graph of $y = -f(x)$ is the image of the graph of $y = f(x)$ after it has been reflected in the x-axis.

The graph of $y = f(-x)$ is the image of the graph of $y = f(x)$ after it has been reflected in the y-axis.

An **even** function is such that:
$$f(x) = f(-x)$$
so its graph is symmetrical about the y-axis.

An **odd** function is such that:
$$f(-x) = -f(x)$$
so its graph has rotational symmetry of order 2 about the origin.

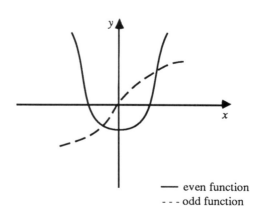

— even function
- - - odd function

■ $f(x) = \dfrac{1}{\sqrt{(x+3)}}$

Sketch the graph of f(x) and state its domain and range.

●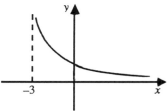

Domain is all real numbers greater than −3.
Range is all positive real numbers.

■ If $f(x) = \dfrac{3-x}{x+2}$, find $f^{-1}(x)$.

● The flow diagram method cannot be used here.

First write y instead of f(x) and make x the subject of the formula.

$y = \dfrac{3-x}{x+2} \Rightarrow y(x+2) = 3-x$

$\begin{aligned} xy + 2y &= 3 - x \\ xy + x &= 3 - 2y \\ x(y+1) &= 3 - 2y \\ x &= \dfrac{3-2y}{y+1} \end{aligned}$

So $f^{-1}(x) = \dfrac{3-2x}{x+1}$

■ If fg(x) = sin 2x, identify possible f(x) and g(x) and write down gf(x).

● f(x) = sin x and g(x) = 2x
gf(x) = 2 sin x

■ The function f(x) has the graph as shown.

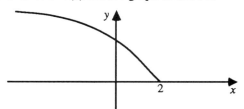

(a) Sketch the graph of f(x + 2).

(b) Sketch the graph of f(− x).

●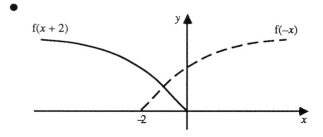

---

1 (a) f(x) = x³ + 1 and g(x) = cos x

    (i) Find the value of gf(−1).

    (ii) Write down a formula for gf(x).

  (b) $f(x) = \dfrac{3-x}{x+1}$ and $g(x) = \dfrac{1}{x}$

    Find fg(x) and state its domain and range.

2 Find $f^{-1}(x)$ if:

  (a) f(x) = 3x² + 4

  (b) f(x) = (x + 2)³

  (c) $f(x) = \dfrac{2x+1}{5x-3}$

3 (a) Make x the subject of $y = \dfrac{a}{x+b}$.

  (b) Re-arrange the formula $t = \dfrac{a}{a-b}$

    (i) to express b in terms of a and t;

    (ii) to express a in terms of b and t.

4 The formula $C = \dfrac{N}{20} + 9$ gives the cost in pounds of N units of telephone use, at 5 pence a unit, when there is a standing charge of £9.

  (a) Rearrange this formula to show the number (N) of units obtained at a cost of C pounds.

  (b) Give the formula for C, the cost of N units at x pence each, when the standing charge is y pounds.

5 The graph of f(x) is shown below.

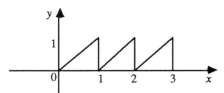

Sketch the graphs of:

  (a) −f(x)   (b) f(−x)

  (c) $f^{-1}(x)$ (d) f(x + 1)

Trigonometric functions

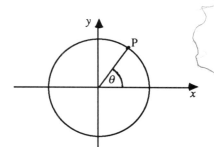

If the circle has unit radius, the coordinates of point P are (cos θ, sin θ).

$$\tan \theta = \frac{\sin \theta}{\cos \theta}$$

Graphs

The graphs of sin θ and cos θ have period 360° and amplitude 1. The graph of tan θ has period 180°.

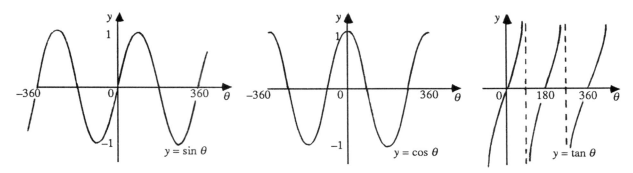

Sin θ and tan θ are odd functions, so sin $(-\theta) = -$ sin θ and tan $(-\theta) = -$ tan θ.
Cos θ is an even function, so cos $(-\theta) =$ cos θ.

From the graph of $y = $ sin $\theta°$, the graph of $y = a$ sin $(b\theta + c)° + d$ is obtained by a stretch of scale factor a parallel to the y-axis, a stretch of scale factor $\frac{1}{b}$ parallel to the x-axis and a translation of $\begin{bmatrix} -\frac{c}{b} \\ d \end{bmatrix}$.

$-\frac{c}{b}$ is called the **phase shift** from $y = $ sin $\theta°$ to $y = a$ sin $(b\theta + c)°$ and the new graph has amplitude a and period $\frac{360°}{b}$.

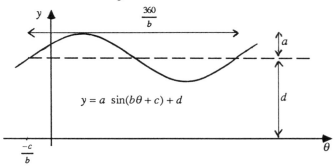

$$y = a \ \sin(b\theta + c) + d$$

Inverse functions

If sin $a = x$, then $a = $ sin$^{-1} x$.

sin$^{-1} x$, cos$^{-1} x$ and tan$^{-1} x$ are only functions if their ranges are restricted, for example, to the sets of **principal values**:

$$-90° < \sin^{-1} x \le 90° \qquad 0 \le \cos^{-1} x < 180° \qquad -90° \le \tan^{-1} x \le 90°$$

Equations

A trigonometric equation of the form a sin $(bx + c)° = k$ will have an infinite number of solutions, two of which will generally be in the range:

$$0 \le x \le \frac{360}{b}$$

■ Sketch the graph of $y = 3 \sin (4t - 100)°$.

● The graph will have amplitude 3, period $\frac{360°}{4}$ and have a phase shift of $\frac{100°}{4}$.

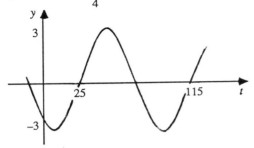

■ Solve the equation $\sin \theta = 0.5$ for $0 \le \theta \le 720°$.

● The principal value of $\sin^{-1} 0.5$ is $30°$.

From the graph of $y = \sin \theta$ it is clear that a second solution is $180° - 30° = 150°$.

Since the period of $y = \sin \theta$ is $360°$ the two other solutions required are:

$$360° + 30° = 390° \text{ and } 360° + 150° = 510°$$

■ Solve the equation

$$2 \cos (3t + 20)° = 1.5$$

for $0 \le t \le 180°$.

● $\cos (3t + 20)° = 0.75$

If $0 \le t \le 180$, then $20 \le 3t + 20 \le 560$

Values of $\cos^{-1} 0.75$ in the required range are:

$$41.4, \ 360 - 41.4, \ 360 + 41.4$$
$$3t + 20 = 41.4, 318.6, 401.4$$
$$t = 7.1°, 99.5°, 127.1°$$

■ The height in metres of water, t hours after high tide, is given by $h = 0.8 \cos 30t° + 4$.

How long is it between successive high tides and what is the minimum height of water?

● The period of $\cos 30t$ is $\frac{360}{30}$, so there are 12 hours between successive high tides.

The minimum value of $\cos 30t$ is -1, so the minimum height is $-0.8 + 4 = 3.2$m.

---

1 Suggest equations for these graphs.

(a)

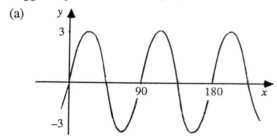

(b)

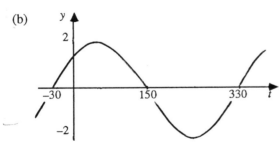

2 Give three other angles in the range $-360° \le x \le 360°$ which have the same cosine as $54°$.

3 $y = 3 \cos (2x - 30)°$

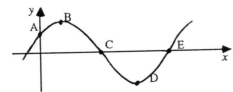

Write down the coordinates of A, B, C, D and E.

4 Solve $3 \sin x = 2$ for $0 \le x \le 900°$ and use the solutions to solve the equation:

$$3 \sin (4t + 30)° = 2 \text{ for } 0 \le t \le 180°$$

5 $2 \sin x = 3 \cos x$

Write down the value of $\tan x$ and hence solve the equation for $-180° \le x \le 180°$.

6

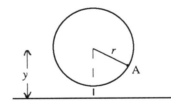

A is a chair on a fairground wheel, which starts at its lowest point. Its height (h metres) above the ground t seconds later is given by:

$$h = 7.5 - 6 \cos 8t°$$

(a) Write down the lengths y metres and r metres.

(b) Find the value of $\cos 8t°$ when A is 6 metres above the ground.

(c) Work out the first two times at which A is 6 metres above the ground.

25

Exponential growth

Growth is **exponential** when there is a constant, called the **growth factor**, such that during each unit time interval the amount present is multiplied by this factor.

The general growth function has an equation of the form

$$y = ka^x$$

where a is the growth factor and k is the value of y when $x = 0$.

Laws of indices

For any positive number a and any numbers p and q:

$$a^0 = 1 \qquad a^{-p} = \frac{1}{a^p}$$

$$a^p \times a^q = a^{p+q} \qquad a^p \div a^q = a^{p-q} \qquad (a^p)^q = a^{pq}$$

If n is non-zero, $a^{\frac{1}{n}} = \sqrt[n]{a}$.

Logarithms

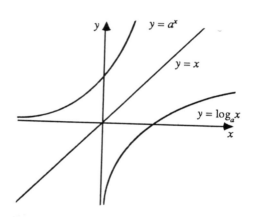

If $f(x) = a^x$, then $f^{-1}(x) = \log_a x$.

Or, $y = a^x \implies x = \log_a y$.

Laws of logarithms

$$\log_a 1 = 0$$
$$\log_a a = 1$$
$$\log_a mn = \log_a m + \log_a n$$
$$\log_a a^x = x \log_a a = x$$

$$\log_a \frac{1}{a} = -1$$
$$\log_a \frac{m}{n} = \log_a m - \log_a n$$
$$a^{(\log_a x)} = x$$

$\log_{10} x$ is generally written just as $\log x$.

$a^x = b$

Equations of the form $a^x = b$ can be solved using logarithms.

■ Solve $3^x = 9.7$.

●
$$3^x = 9.7$$
$$\implies \log 3^x = \log 9.7$$
$$x \log 3 = \log 9.7$$
$$x = \frac{\log 9.7}{\log 3}$$

$$x = 2.07 \text{ (to 2 d.p.)}$$

■ (a) Simplify $(a^4 \times a^5 \div a^6)^3$.

(b) Evaluate: (i) $\log_2 32$

(ii) $\log_9 3$

● (a) $(a^{4+5-6})^3 = (a^3)^3 = a^9$

(b) (i) $2^5 = 32$, so $\log_2 32 = 5$

(ii) $9^{\frac{1}{2}} = 3$, so $\log_9 3 = \frac{1}{2}$

■ The number of bacteria in a colony was initially 400 and then increased by 60% each hour. Explain why the number of bacteria t hours later was 400×1.6^t. After how long were there 7000 bacteria?

● An increase of 60% each hour means the growth factor is 1.6, so after t hours the number is multiplied by 1.6^t.

$400 \times 1.6^t = 7000 \Rightarrow 1.6^t = 17.5$

$t = \dfrac{\log 17.5}{\log 1.6} = 6.1$ (to 2 s.f.)

There were 7000 bacteria after just over 6 hours.

■ The population of greenfly on a plant grows from 15 to 350 during 24 hours. Assuming that the growth can be modelled by $P = ka^t$, calculate k and a.

● $15 = ka^0$ ①

$350 = ka^{24}$

From ① , $k = 15$

So $350 = 15a^{24}$

$a^{24} = \dfrac{350}{15}$

$a = \left(\dfrac{350}{15}\right)^{\frac{1}{24}} = 1.14$ (to 3 s.f.)

■ The population of a village was 110, 130, 156, 188 and 225 in 1950, 1960, 1970, 1980 and 1990 respectively. Was the population growth exponential? Justify your answer.

● For exponential growth, the growth factors for each 10-year period should be equal.

$\dfrac{130}{110}$, $\dfrac{156}{130}$, $\dfrac{188}{156}$ and $\dfrac{225}{188}$ are all near to 1.2, so the growth is approximately exponential.

---

1 Evaluate:

(a) $\log_3 \dfrac{1}{9}$ (b) $\log_5 0.2$

(c) $\log_9 27$ (d) $\log_2 \sqrt{8}$

2 Solve the equations:

(a) $2^{(5t+1)} = 64$

(b) $5^{(t-2)} = 12$

3. Sketch the graphs of $\log x$ and $\log 3x$ and describe the transformation which maps the graph of $\log x$ onto that of $\log 3x$.

4 A collection of villages is designated as a new town. In the first year the population grows from 14 000 to 17 500. If the growth of population is exponential, what is the population, to the nearest 500, at the end of the third year?

5 The proportion of the population who do not have the use of a car fell from 72% in 1961 to 37% in 1987. Assuming that this can be modelled by the equation $P = ka^t$, where t is the numbers of years after 1961:

(a) calculate k and a;

(b) estimate the first year in which at least 75% of all households have the use of a car.

6 The number of bacteria in a colony is initially 400. After t hours there are 400×1.7^t.

(a) How many will there be after 12 hours?

(b) How long does it take for the number of bacteria to double?

(c) At a time T, the number of bacteria is one million times greater than the number of hours of growth.

Show that: $T = \dfrac{\log (2500\,T)}{\log 1.7}$

(d) Use the iterative formula

$T_{n+1} = \dfrac{\log (2500\,T_n)}{\log 1.7}$

to find the value of T.

7 An article cost £100 in January 1980, but by January 1992 its price had risen to £370.

(a) What was the growth factor in the price over the 12-year period?

(b) Assuming that the growth in price was exponential:

(i) calculate the annual growth factor;

(ii) work out the price in January 1985.

Definition of a radian

A **radian** is the angle subtended by an arc equal in length to the **radius** of a circle.

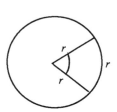

Since the circumference of a circle is $2\pi r$, there are 2π radians in a complete turn.

Graphs

2π radians = 360°; so, if radians are used, the period of the graph of $y = \sin ax$ is $\dfrac{2\pi}{a}$.

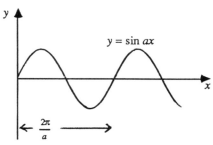

Arc length and sector area

If θ is measured in radians:

Arc length of sector, $s = r\theta$

Area of sector $= \dfrac{1}{2} r^2\theta$

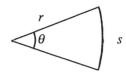

Converting radians ↔ degrees

2π radians is written as $2\pi^c$ (c stands for 'circular measure').

Since $2\pi^c = 360°$, $\pi^c = 180°$ and $\dfrac{\pi^c}{2} = 90°$

$$1^c = \frac{180°}{\pi} \approx 57.3°$$

In practice the 'c' is often left out and it is assumed that, for example,

$\sin 2$ means $\sin 2^c$.

Differentiation and integration

Provided x is measured in radians:

$\dfrac{\mathrm{d}}{\mathrm{d}x} (\sin x) = \cos x$

$\dfrac{\mathrm{d}}{\mathrm{d}x} (\sin ax) = a \cos ax$

$\dfrac{\mathrm{d}}{\mathrm{d}x} (\cos x) = -\sin x$

$\dfrac{\mathrm{d}}{\mathrm{d}x} (\cos ax) = -a \sin ax$

$\displaystyle\int \sin x \,\mathrm{d}x = -\cos x + c$

$\displaystyle\int \cos x \,\mathrm{d}x = \sin x + c$

$\displaystyle\int \sin ax \,\mathrm{d}x = -\dfrac{1}{a} \cos ax + c$

$\displaystyle\int \cos ax \,\mathrm{d}x = \dfrac{1}{a} \sin ax + c$

■ The area of this sector is 25 cm². Calculate θ.

6 cm

● Since area of sector = $\frac{1}{2} r^2 \theta$:

$\frac{1}{2}$ x 36 x θ = 25

$\theta = \frac{25}{18} = 1.4^c$

■ An aeroplane is 'looping the loop' at an airshow. Its height, h metres, at a time t seconds after the start at the bottom of the loop is given by:

$h = 200 - 100 \cos 0.1t$

(a)   What is the plane's maximum height?

(b)   Find the plane's maximum rate of climb.

● (a)   The minimum value of $\cos 0.1t$ is -1, so the maximum height is $200 - (-100) = 300$m.

(b)   Rate of climb = $\frac{dh}{dt}$

$\frac{dh}{dt} = 10 \sin (0.1t)$

The maximum rate of climb is 10 m s⁻¹.

■ Change 3ᶜ to degrees.

● $\pi^c = 180°$, so $1^c = \frac{180°}{\pi}$

$3^c = \frac{540°}{\pi} = 171.9°$ (to 1 d.p.)

■ Find the turning points in the range $0 \le x \le 2\pi$ for $y = \sin 2x + \cos 2x$.

● $\frac{dy}{dx} = 2 \cos 2x - 2 \sin 2x$

$\frac{dy}{dx} = 0 \Rightarrow \sin 2x = \cos 2x$

So $\tan 2x = 1$

At the required turning points,

$2x = \tan^{-1} 1$, for $0 \le 2x \le 4\pi$

$2x = \frac{\pi}{4}, \frac{5\pi}{4}, \frac{9\pi}{4}$ and $\frac{13\pi}{4}$

The turning points are:

$(\frac{\pi}{8}, 1.414), (\frac{5\pi}{8}, -1.414), (\frac{9\pi}{8}, 1.414)$ and

$(\frac{13\pi}{8}, -1.414)$

---

1   Change to degrees:

(a) $\frac{\pi}{3}^c$ (b) $\frac{7\pi}{8}^c$ (c)   0.3ᶜ

2   Change to radians, giving the answer in terms of π where appropriate:

(a)   45°   (b)   100°   (c)   225°   (d)   280°

3   A gardener designs a flower bed in the shape of a sector of a circle of radius 4 m and angle 1.4^c.

(a)   He plants petunias, which need 300 cm² of ground each. How many can he plant?

(b)   He surrounds the bed with an edging strip. How long a strip does he need?

4   A ship's motion in heavy seas is modelled by $h = 10 \sin \frac{\pi t}{5}$, where h metres is its displacement from normal position and time t is measured in seconds.

(a)   What is the amplitude and period of the motion?

(b)   Crests of surrounding waves are above deck level if $h \le -5$. For how long in each cycle does this occur?

(c)   A passenger feels sea-sick when they are descending at more than 4 m s⁻¹. For what percentage of the cycle does this occur?

5   Give, to 3 s.f., two values of t for which $\tan t = \pi$.

6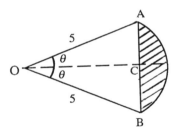

(a)   What, in terms of θ, is the length of the arc AB?

(b)   What, in terms of θ, is the length of AC?

(c)   Hence, or otherwise, show that the perimeter of the shaded area is given by $p = 10(\theta + \sin \theta)$.

(d)   Find the value of $\frac{dp}{d\theta}$ when $\theta = 0$.

7   A pastry cutter is made in the shape of a sector of a circle of radius r cm and with perimeter 42 cm.

(a)   Find θ and show that the area to be cut out is given by $A = 21r - r^2$.

(b)   Find $\frac{dA}{dr}$ and so calculate the maximum area which such a pastry cutter could cut out.

29

Definition of e

If $y = a^x$, $\dfrac{dy}{dx} = k \times a^x$, where a and k are constants.

So, for example:

$$\frac{d}{dx}\,(2^x) \approx 0.69 \times 2^x \qquad\qquad \frac{d}{dx}\,(3^x) \approx 1.1 \times 3^x$$

e is the number such that $\dfrac{d}{dx}\,(e^x) = e^x$.

e is an irrational number. $e = 2.718$ (to 3 d.p.)

e^x is sometimes written as exp (x).

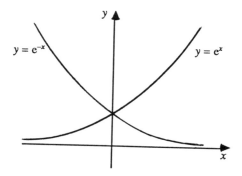

Differentiation and integration

Since $\dfrac{d}{dx}\,(e^x) = e^x$, $\displaystyle\int e^x\,dx = e^x + c$

$\dfrac{d}{dx}\,(e^{ax}) = a\,e^{ax}$ and so $\displaystyle\int e^{ax}\,dx = \dfrac{1}{a}\,e^{ax} + c$

ln x

$\ln x = \log_e x$, so $\ln x$ is the inverse of the function e^x.

The graph of $y = \ln x$ can be obtained by reflecting the graph of $y = e^x$ in the line $y = x$.

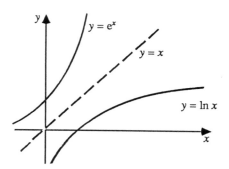

$$\frac{d}{dx}\,(\ln x) = \frac{1}{x}$$

$\ln ax = \ln a + \ln x$, where $\ln a$ is constant, and so

$$\frac{d}{dx}\,(\ln ax) \;=\; \frac{d}{dx}\,(\ln x) \;=\; \frac{1}{x}$$

$\displaystyle\int \frac{1}{x}\,dx$ can be written as $\ln x + c$ or as $\ln ax$.

Equations

$y = e^x \Rightarrow x = \ln y$

■ Solve $e^x = 5$

● $\qquad e^x = 5 \Rightarrow \ln 5 = x$

$\qquad\quad x = 1.61$ (to 2 d.p.)

■ Find the equation of the tangent to the curve $y = 3e^{2x}$ at $x = 1$.

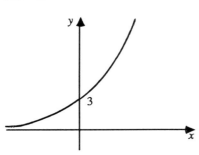

● $\dfrac{dy}{dx} = 6e^{2x}$

At $x = 1$:

$y = 3e^2 = 22.2$ (to 1 d.p.)

$\dfrac{dy}{dx} = 6e^x = 44.3$ (to 1 d.p.)

The equation of the tangent is:

$\dfrac{(y - 22.2)}{(x - 1)} = 44.3$

$y - 22.2 = 44.3\,x - 44.3$

$y = 44.3\,x - 22.1$

■ Find $\dfrac{dy}{dx}$ if $y = \dfrac{1}{e^{2x}}$

● $\dfrac{1}{e^{2x}} = e^{-2x}$

$\dfrac{dy}{dx} = -2e^{-2x} = -\dfrac{2}{e^{2x}}$

■ A cup of coffee cools so that its temperature is given by $T = 12 + 80e^{-t/6}$, where T is the temperature in °C and t the time in minutes.

(a) What is its initial temperature?

(b) How long does it take to cool to 50°C?

● (a) The initial temperature is when $t = 0$.

So $T = 12 + 80e^0 = 92$°C

(b) $50 = 12 + 80e^{-t/6} \Rightarrow e^{-t/6} = 0.475$

$-\dfrac{1}{6}t = \ln 0.475$

$t = \ln 0.475 \times (-6) = 4.47$

The coffee cools to 50°C in just under $4\frac{1}{2}$ minutes.

---

1 Match up equations (i), (ii) and (iii) to the graphs A, B and C.

(i) $y = 3e^{-x}$ (ii) $y = 3(1 - e^{-2x})$

(iii) $y = \ln 3x$

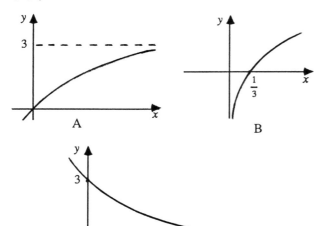

2 Write down the values of :

(a) $\ln e^3$ (b) $\ln \sqrt{e}$ (c) $\ln \dfrac{1}{e}$

3 Evaluate: $\displaystyle\int_1^2 e^{3x}\,dx$

4 An isolated boarding school has 500 students. Two sixth-formers start a rumour which spreads round the school. The number who have heard the rumour t hours after it began is given by:

$$N = \dfrac{500}{1 + 249e^{-t}}$$

(a) How many students have heard the rumour after 1 hour?

(b) How long does it take for half the students to have heard the rumour?

(c) How long is it before only one student does not know the rumour?

5 Find the equation of the tangent to the graph of $y = \ln x$ at $x = 0.5$.

6 The mass, m kg, of radioactive lead remaining in a sample t hours after observation begins is given by the equation $m = 3e^{-0.4t}$.

(a) Find the mass left after 10 hours.

(b) Find how long it takes for the mass to fall to half its initial value.

(c) Find the rate at which the mass is decaying after 4 hours.

Transformations and graphs

Some changes to f(x) lead to simple geometric transformations of the graph of y = f(x):

Change in function	Transformation
Replace x with $-x$	Reflection in the y-axis
Replace y with $-y$	Reflection in the x-axis
Interchange x and y	Reflection in $y = x$
Replace x with kx	One way stretch from y-axis scale factor $\frac{1}{k}$
Replace y with ky	One way stretch from x-axis scale factor $\frac{1}{k}$
Replace x with $x + k$	Translation with vector $\begin{bmatrix} -k \\ 0 \end{bmatrix}$
Replace y with $y + k$	Translation with vector $\begin{bmatrix} 0 \\ -k \end{bmatrix}$

Combined transformations

Consider the effect of replacing x by $3x$ and y by $y + 2$ on the graph of $y = x^2$.

$$y + 2 = (3x)^2$$
$$\Rightarrow \quad y = 9x^2 - 2$$

The graph of $y = x^2$ is stretched with scale factor $\frac{1}{3}$ from the y-axis and translated by $\begin{bmatrix} 0 \\ -2 \end{bmatrix}$.

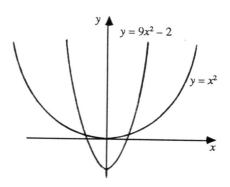

Circles and ellipses

$x^2 + y^2 = 1$ is the equation of the circle of radius 1 and centre (0, 0).

The equation can be rewritten as $y = \pm \sqrt{(1 - x^2)}$, so plotting $y = \sqrt{(1 - x^2)}$ and $y = -\sqrt{(1 - x^2)}$ gives the complete circle.

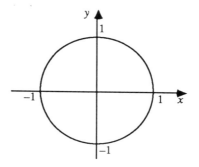

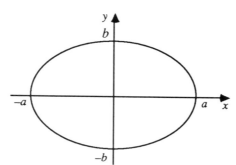

$$\left(\frac{1}{a}x \right)^2 + \left(\frac{1}{b}y \right)^2 = 1 \qquad (a > b)$$

is the equation of an ellipse, centre (0, 0), with major axis of length $2a$ and minor axis of length $2b$.

■ Graph A has equation $y = 2x^3$. Graph B is obtained from graph A by a translation which takes the origin to the point $(-1, 2)$. Sketch the two graphs and work out the equation of graph B.

●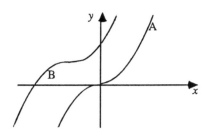

The equation of graph B is $y = 2(x + 1)^3 + 2$.

■ The graph of $y = e^x$ is mapped onto that of $y = 8 + e^{x+3}$ by the translation $\begin{bmatrix} r \\ s \end{bmatrix}$. State the values of r and s.

● $y - 8 = e^{x+3}$, so y has been replaced by $y - 8$ and x by $x + 3$. So $r = -3$ and $s = 8$.

■ The graph of $y = x^2$ is translated with vector $\begin{bmatrix} 1 \\ 0 \end{bmatrix}$ and then stretched with scale factor $\frac{1}{2}$ from the y-axis. Sketch the resulting graph and work out its equation.

●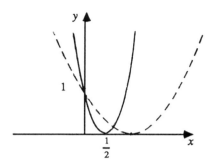

In the equation $y = x^2$, replace x by $x - 1$, then replace x by $2x$.

The new equation is $y = (2x - 1)^2$.

---

1   In $0.5x = \ln x + \ln 0.5$, so the graph of $y = \ln x$ can be mapped onto the graph of $y = \ln 0.5x$ by a translation. The mapping could also be described as a stretch. Describe these two transformations fully.

2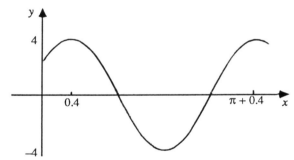

Give an equation for the graph above:

(a)   in the form $y = a \cos (bx + c)$;

(b)   in the form $y = a \sin (bx + c)$.

3   Find the image of $y = \dfrac{1}{x}$

(a)   after a translation through $\begin{bmatrix} 5 \\ -9 \end{bmatrix}$ followed by reflection in the x-axis;

(b)   after a reflection in the y-axis followed by a translation of $\begin{bmatrix} 2 \\ -3 \end{bmatrix}$.

Give your answers in the form $y = \dfrac{ax + b}{cx + d}$.

4   The graph of $y = x^2$ is stretched with scale factor $\frac{1}{2}$ from the y-axis and then translated with vector $\begin{bmatrix} 1 \\ 0 \end{bmatrix}$.

Sketch the resulting graph and work out its equation. (Notice that the answer is **not** the same as in the worked example above!)

5   Suggest a suitable equation for this graph .

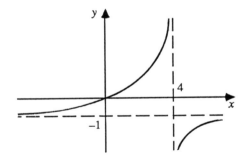

6   (a)   An ellipse cuts the axes at the points $(-3, 0)$, $(0, 2)$, $(3, 0)$ and $(0, -2)$. Write down its equation in the form

$$\frac{x^2}{a^2} + \frac{y^2}{b^2} = 1$$

(b)   Find the equation of the ellipse, with centre $(2, 1)$, whose major axis is parallel to the x-axis and of length 8 units and whose minor axis is parallel to the y-axis and of length 6 units.

MATHEMATICAL METHODS

Pythagoras in 2D and 3D

$h^2 = a^2 + b^2$

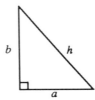

Extending into three dimensions:

$d^2 = a^2 + b^2 + c^2$

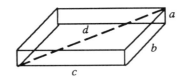

The equation of a circle of radius r about the point (a, b) is

$$(x - a)^2 + (y - b)^2 = r^2$$

Extending into three dimensions:

The equation of a sphere of radius r about the point (a, b, c) is

$$(x - a)^2 + (y - b)^2 + (z - c)^2 = r^2$$

Identities

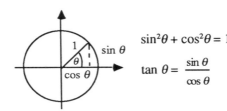

$\sin^2\theta + \cos^2\theta = 1$

$\tan\theta = \dfrac{\sin\theta}{\cos\theta}$

■ Solve $\sin x = 3\cos x$ for $0 \le x \le 2\pi$.

● The graphs of $y = \sin x$ and $y = 3\cos x$ indicate two solutions.

$\tan x = 3$ (dividing through by $\cos x$)
$x = 1.25$ or 4.39 in the range $0 \le x \le 2\pi$

$r\sin(\theta + \alpha)$

$a\sin\theta + b\cos\theta = r\sin(\theta + \alpha)$ where r and α can be found from the triangle

$r = \sqrt{(a^2 + b^2)}$

and $\alpha = \tan^{-1}\dfrac{b}{a}$

$r\sin(\theta + \alpha)$ is a sine wave, amplitude r, phase shifted by α in the negative x-direction.

Addition and double angle formulas

$\sin(A + B) = \sin A\cos B + \cos A\sin B$
$\sin(A - B) = \sin A\cos B - \cos A\sin B$

replacing B by A gives $\sin 2A = 2\sin A\cos A$

$\cos(A + B) = \cos A\cos B - \sin A\sin B$
$\cos(A - B) = \cos A\cos B + \sin A\sin B$

replacing B by A gives $\cos 2A = \cos^2 A - \sin^2 A$
$= 2\cos^2 A - 1$
$= 1 - 2\sin^2 A$

Cosine and sine rules

If a triangle is **not** right-angled, lengths and angles can be calculated using the sine and cosine rules.

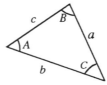

Cosine rule:

$a^2 = b^2 + c^2 - 2bc\cos A$
$b^2 = a^2 + c^2 - 2ac\cos B$
$c^2 = a^2 + b^2 - 2ab\cos C$

Sine rule:

$$\dfrac{a}{\sin A} = \dfrac{b}{\sin B} = \dfrac{c}{\sin C}$$

The area of the triangle is $\dfrac{1}{2}ab\sin C$

■ Explain why $\sin \theta = \cos(90° - \theta)$ by

  (a) using a diagram;

  (b) using the addition formulas.

● (a)

If $C = \theta$, then $A = 90° - \theta$

so $\sin \theta = \cos(90° - \theta) = \dfrac{c}{b}$

  (b) $\cos(90° - \theta) = \cos 90° \cos \theta + \sin 90° \sin \theta$
$$= 0 \times \cos \theta + 1 \times \sin \theta$$
$$= \sin \theta$$

■ Solve the equation
$\sin x - 2 \cos^2 x = -1$ for $0 \le x \le 2\pi$

● $\cos^2 x = 1 - \sin^2 x$

So $\sin x - 2(1 - \sin^2 x) = -1$
$$\sin x - 2 + 2\sin^2 x = -1$$
or $2\sin^2 x + \sin x - 1 = 0$
$$(2\sin x - 1)(\sin x + 1) = 0$$

$\sin x = 0.5 \Rightarrow x = 0.524, 2.618$
$\sin x = -1 \Rightarrow x = \dfrac{3\pi}{2}$ (or 4.71)

■ Solve $3 \sin x + 4 \cos x = 1$ for $0 \le x \le 180°$.

● Let $3 \sin x + 4 \cos x = r \sin(x + \alpha)$

From the triangle,

$r = \sqrt{(3^2 + 4^2)} = 5$

and $\alpha = \tan^{-1} \dfrac{4}{3} = 53.1°$

So $3 \sin x + 4 \cos x = 1 \Rightarrow 5 \sin(x + 53.1°) = 1$
$$\Rightarrow \sin(x + 53.1°) = 0.2$$
Putting $A = x + 53.1°$
$\sin A = 0.2$ and $A = 11.5°, 168.5°, \ldots$
$\Rightarrow x = 168.5 - 53.1°$
$$= 115.4° \text{ for the range } 0 \le x \le 180°$$

■ For triangle PQR, calculate angle P and use it to find the area of the triangle.

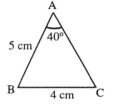

● $9^2 = 8^2 + 5^2 - 2 \times 8 \times 5 \times \cos P$

$\cos P = \dfrac{8^2 + 5^2 - 9^2}{2 \times 8 \times 5}$

$P = 84.3°$

and area $= \dfrac{1}{2} \times 8 \times 5 \times \sin 84.3°$
$$= 19.9 \text{ cm}^2$$

---

1 Use the triangles below to write down exact values of the sine, cosine and tangent of 30°, 45° and 60°.

2 (a) Write down the equation of the circle with centre $(4, 3)$ and radius 5 and find where it intersects the x and y axes.

    (b) Write down the equation of the sphere with centre $(1, -2, -3)$ and radius 5.

3 Solve for $0 \le x \le 2\pi$:

  (a) $2 \cos^2 x + 3 \sin x = 0$

  (b) $\sin 2x = \cos x$

4 (a) Express $3 \sin \theta + 5 \cos \theta$ in the form $r \sin(\theta + \alpha)$ and hence sketch the curve.

    (b) Solve the equation $3 \sin \theta + 5 \cos \theta = 4$ for $0 \le \theta \le 360°$.

5 (a) Expand $\cos(2A + A)$.

    (b) Hence, or otherwise, prove that $\cos 3A = 4 \cos^3 A - 3 \cos A$.

6 (a) Use the sine rule to evaluate $\sin C$.

    (b) Explain why there are two possible values of C.

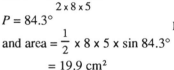

    (c) What happens if BC is 6 cm and the other measurements stay the same?

7 York lies 68 km from Sheffield on a bearing of 021°. Manchester lies 53 km from Sheffield on a bearing of 281°.

Calculate

  (a) the direct distance from York to Manchester;

  (b) the bearing of Manchester from York.

[SMP 1988]

Position vectors

A point may be described in terms of its **coordinates** (x, y) or in terms of its **position vector** $\begin{bmatrix} x \\ y \end{bmatrix}$, which describes a translation from the origin to the point.

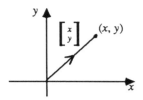

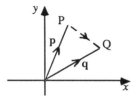

For two points P and Q with position vectors **p** and **q**, the vector describing a translation from P to Q is given by

$$\overrightarrow{PQ} = \mathbf{q} - \mathbf{p}$$

Vector equations of lines

The **vector equation** of a line through a point with position vector **a** and in the direction **b** is

$$\mathbf{r} = \mathbf{a} + \lambda\mathbf{b}$$

In two dimensions,

$$\begin{bmatrix} x \\ y \end{bmatrix} = \begin{bmatrix} a_1 \\ a_2 \end{bmatrix} + \lambda \begin{bmatrix} b_1 \\ b_2 \end{bmatrix}$$

and extending to three dimensions,

$$\begin{bmatrix} x \\ y \\ z \end{bmatrix} = \begin{bmatrix} a_1 \\ a_2 \\ a_3 \end{bmatrix} + \lambda \begin{bmatrix} b_1 \\ b_2 \\ b_3 \end{bmatrix}$$

■ Find a vector equation of the line through the two points A(2, 1) and B(5, –7).

● The direction of the line is given by the vector

$$\overrightarrow{AB} = \begin{bmatrix} 5 \\ -7 \end{bmatrix} - \begin{bmatrix} 2 \\ 1 \end{bmatrix} = \begin{bmatrix} 3 \\ -8 \end{bmatrix}$$

So an equation of the line is

$$\begin{bmatrix} x \\ y \end{bmatrix} = \begin{bmatrix} 2 \\ 1 \end{bmatrix} + \lambda \begin{bmatrix} 3 \\ -8 \end{bmatrix}$$

Scalar product

The **scalar product** of $\mathbf{a} = \begin{bmatrix} a_1 \\ a_2 \\ a_3 \end{bmatrix}$ and $\mathbf{b} = \begin{bmatrix} b_1 \\ b_2 \\ b_3 \end{bmatrix}$ is defined as

$$\mathbf{a.b} = a_1b_1 + a_2b_2 + a_3b_3 = ab \cos \theta$$

where a and b are the magnitudes of the two vectors and θ is the angle between them. To find the angle between vectors **a** and **b**, use

$$\cos \theta = \frac{\mathbf{a.b}}{ab} = \frac{a_1b_1 + a_2b_2 + a_3b_3}{ab}$$

Properties of the scalar product:
$\mathbf{a.b} = \mathbf{b.a}$ and $\mathbf{a.(b + c)} = \mathbf{a.b} + \mathbf{a.c}$
$\mathbf{a.a} = a^2$
$\mathbf{a.b} = 0 \Rightarrow \mathbf{a}$ is perpendicular to **b** or $\mathbf{a} = 0$ or $\mathbf{b} = 0$

Planes

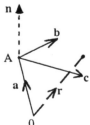

The **vector equation** of any plane can be written in the form

$$\mathbf{r} = \mathbf{a} + \lambda\mathbf{b} + \mu\mathbf{c}$$

where **r** is a general point on the plane, **a** is a particular point on the plane and **b** and **c** are two independent vectors parallel to the plane.

This equation can be written as $\mathbf{r.n} = \mathbf{a.n}$ where **n** is a **normal vector** to the plane [$\mathbf{b.n} = 0$ and $\mathbf{c.n} = 0$] and leads to the **Cartesian equation** of a plane.

The plane with Cartesian equation $ax + by + cz = d$ has normal vector $\begin{bmatrix} a \\ b \\ c \end{bmatrix}$.

The angle between two planes is equal to the angle between the normal vectors of the two planes: $\mathbf{n_1.n_2} = n_1 n_2 \cos \theta$

If the angle between a line and the normal to a plane is β, then the angle between the line and the plane is $90 - \beta$.

■ Determine the value of a for which the lines

$$\begin{bmatrix} x \\ y \\ z \end{bmatrix} = \begin{bmatrix} 1 \\ 2 \\ 4 \end{bmatrix} + s \begin{bmatrix} 3 \\ -1 \\ 2 \end{bmatrix} \text{ and } \begin{bmatrix} x \\ y \\ z \end{bmatrix} = \begin{bmatrix} -2 \\ -5 \\ 8 \end{bmatrix} + t \begin{bmatrix} 0 \\ 4 \\ a \end{bmatrix}$$

intersect. Find this point of intersection.

● Equating the x and y components of the vectors gives

$$1 + 3s = -2 + 0t$$
$$\text{and } 2 - s = -5 + 4t$$

So $s = -1$ and $t = 2$

If the lines are to intersect, the z components must also be equal, i.e. $4 + 2s = 8 + at$
$$\Rightarrow 4 - 2 = 8 + 2a \Rightarrow a = -3$$

so the lines intersect where

$$\begin{bmatrix} x \\ y \\ z \end{bmatrix} = \begin{bmatrix} 1 \\ 2 \\ 4 \end{bmatrix} - 1\begin{bmatrix} 3 \\ -1 \\ 2 \end{bmatrix} = \begin{bmatrix} -2 \\ 3 \\ 2 \end{bmatrix}$$

giving a point of intersection at $(-2, 3, 2)$.

■ Find the angle between the vectors $\begin{bmatrix} 2 \\ -1 \\ 3 \end{bmatrix}$ and $\begin{bmatrix} 4 \\ 3 \\ 2 \end{bmatrix}$.

● $\mathbf{a.b} = 2 \times 4 + (-1) \times 3 + 3 \times 2 = 11$
$a = \sqrt{(4 + 1 + 9)} = \sqrt{14}$ $b = \sqrt{(16 + 9 + 4)} = \sqrt{29}$

So $\cos\theta = \dfrac{11}{\sqrt{14}\sqrt{29}} \Rightarrow \theta = 56.9°$

■ Find the equation of the plane which is perpdendicular to the line $\mathbf{r} = \begin{bmatrix} 2 \\ 5 \\ -4 \end{bmatrix} + \lambda \begin{bmatrix} -3 \\ 2 \\ 1 \end{bmatrix}$ and passes through the point $(-3, 1, 0)$.

● The direction of the line is given by $\begin{bmatrix} -3 \\ 2 \\ 1 \end{bmatrix}$ and this is the normal $\mathbf{n}$ to the plane.

Thus $\mathbf{n.r} = \mathbf{n.a}$ gives $\begin{bmatrix} -3 \\ 2 \\ 1 \end{bmatrix} . \begin{bmatrix} x \\ y \\ z \end{bmatrix} = \begin{bmatrix} -3 \\ 2 \\ 1 \end{bmatrix} . \begin{bmatrix} -3 \\ 1 \\ 0 \end{bmatrix}$

$\Rightarrow -3x + 2y + z = 11$

---

1 Find the vector equation of the line

(a) through the point $(-1, 4)$ parallel to the vector $\begin{bmatrix} 5 \\ -6 \end{bmatrix}$;

(b) joining the points $(-1, 4, 2)$ and $(1, 5, 0)$.

2 A line joins the two points $(0, 1, 1)$ and $(2, 0, 0)$ and a second line joins the point $(3, 0, 1)$ to the origin.

(a) Write down the vector equations of the two lines.

(b) Determine whether the two lines intersect.

3 (a) What can you say about $\mathbf{a}$ and $\mathbf{b}$ if

(i) $\mathbf{a.b} = 0$;   (ii) $\mathbf{a.b} = ab$?

(b) Which pair(s) of the following vectors are

(i) perpendicular;   (ii) parallel?

$$\mathbf{c} = \begin{bmatrix} 5 \\ -3 \\ 2 \end{bmatrix} \quad \mathbf{d} = \begin{bmatrix} 12 \\ 18 \\ -3 \end{bmatrix} \quad \mathbf{e} = \begin{bmatrix} -30 \\ 18 \\ -12 \end{bmatrix}$$

4 Find the equation of the plane through the point $(5, -1, 6)$ and parallel to the vector $\begin{bmatrix} -2 \\ 4 \\ 1 \end{bmatrix}$ and the x-axis

(a) in vector form;

(b) in Cartesian form.

5 Find the angle between

(a) the plane $2x - y = 5$ and the x-axis;

(b) the planes $2x + y = 5$ and $3x + 2z = 7$.

6 Find the point of intersection of the line

$$\begin{bmatrix} x \\ y \\ z \end{bmatrix} = \begin{bmatrix} 2 \\ 1 \\ 0 \end{bmatrix} + \lambda \begin{bmatrix} 1 \\ -4 \\ 3 \end{bmatrix} \text{ with the plane } x - 3y + z = 15.$$

7 A square-based pyramid has vertices at $A(0, 0, 0)$, $B(0, 8, 0)$, $C(8, 8, 0)$, $D(8, 0, 0)$ and $E(4, 4, 6)$. Find

(a) the angle between lines AB and AE;

(b) the angle between lines EA and EC;

(c) the angle between the planes ABCD and ABE;

(d) the angle between the planes ABE and BCE.

8 A laser is positioned at the point $(3, 5, 2)$, directed at the point $(25, 18, 9)$. A flat mirror is placed at a right angle to the laser beam. The mirror passes through the point $(15, 15, 5)$.

(a) Give the equation of the line of the laser beam.

(b) Give the equation of the plane of the mirror.

(c) Give the coordinates of the point where the laser beam hits the mirror.

Binomial expressions

Algebraic expressions which have two terms, for example, $a + b$, $3x - y$, $p^2 + p$, are known as **binomials.**

For small integer values of n, the binomial expression

$$(a + b)^n$$

can be expanded using the nth line of Pascal's triangle.

■ Expand $(x - y)^4$

● $(x - y)^4 = 1x^4 + 4x^3(-y)^1 + 6x^2(-y)^2$
$$+ 4x(-y)^3 + 1(-y)^4$$

$$= x^4 - 4x^3y + 6x^2y^2 - 4xy^3 + y^4$$

The numbers in Pascal's triangle are referred to as **binomial coefficients** using the notation $\binom{n}{r}$, where $\binom{n}{r} = \dfrac{n!}{r!(n-r)!}$

Binomial theorem

For n a positive integer,

$$(a+b)^n = \binom{n}{0}a^n b^0 + \binom{n}{1}a^{n-1}b^1 + \binom{n}{2}a^{n-2}b^2 + \ldots + \binom{n}{r}a^{n-r}b^r + \ldots + \binom{n}{n}a^0 b^n$$

The series is valid for all values of a and b.

Certain values of $\binom{n}{r}$, sometimes referred to as $_nC_r$, can be found in mathematical tables and on many calculators.

Binomial series

The **binomial series** extends the binomial expansion for **all** values of n, provided x is small.

$$(1 + x)^n = 1 + nx + \frac{n(n-1)}{2!}x^2 + \frac{n(n-1)(n-2)}{3!}x^3 + \ldots \quad \text{for} -1 < x < 1$$

Error

Expressing a result in the form $a \pm e$ is a way of stating that the result lies between $a - e$ and $a + e$. The **error** e measures the largest possible difference between the actual value and a.

Measurements, and their errors, are combined in the following ways:

$$(a \pm e) + (b \pm f) = (a + b) \pm (e + f)$$
$$(a \pm e) - (b \pm f) = (a - b) \pm (e + f)$$

$$k(a \pm e) = ka \pm ke$$

$a \pm e$ can be written as $a(1 \pm r)$ where $r = \dfrac{e}{a}$ (called the **relative error**). Measurements in this form are combined in the following ways provided the relative errors are small.

$$a(1 \pm r) \times b(1 \pm s) \approx ab[1 \pm (r + s)]$$

$$\frac{a(1 \pm r)}{b(1 \pm s)} \approx \frac{a}{b}[1 \pm (r + s)]$$

■ Expand: (a) $(2 - 3a)^4$

      (b) $(2 - 3a)^{-4}$

For which values of a are these expansions valid?

● (a) $(2 - 3a)^4 = 2^4 + \binom{4}{1} 2^3 (-3a) + \binom{4}{2} 2^2 (-3a)^2$

                           $+ \binom{4}{3} 2(-3a)^3 + (-3a)^4$

            $= 16 - 96a + 216a^2 - 216a^3 + 81a^4$

This is valid for all values of a.

(b) $(2 - 3a)^{-4} = 2^{-4} (1 - \frac{3}{2} a)^{-4}$

               $= 2^{-4} \left(1 + (-4)\left(\frac{-3}{2}a\right) + \frac{(-4) \times (-5)}{2 \times 1} \left(\frac{-3}{2}a\right)^2 \right.$

                  $\left. + \frac{(-4) \times (-5) \times (-6)}{3 \times 2 \times 1} \left(\frac{-3}{2}a\right)^3 + \dots \right)$

               $= \frac{1}{16} \left(1 + 6a + \frac{45}{2} a^2 + \frac{135}{2} a^3 + \dots \right)$

This is valid for $-1 < \frac{3}{2} a < 1$ or $\frac{-2}{3} < a < \frac{2}{3}$.

■ A rope, length l metres, encloses a triangle.

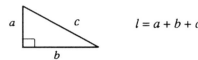

$l = a + b + c$

If $l = 24 \pm 0.5$ m, $b = 8 \pm 0.2$ m and $c = 10 \pm 0.2$ m calculate

(a) the length a;

(b) the area of the triangle.

● (a) $a = l - (b + c)$

         $= (24 \pm 0.5) - ((8 \pm 0.2) + (10 \pm 0.2))$

         $= (24 \pm 0.5) - (18 \pm 0.4)$

         $= 6 \pm 0.9$ m

(b) Area $= \frac{1}{2} ab = \frac{1}{2} (6 \pm 0.9)(8 \pm 0.2)$

            $= \frac{1}{2} \times 6 (1 \pm 0.15) \times 8(1 \pm 0.025)$

            $\approx 24 (1 \pm 0.175)$

            $\approx 24 \pm 4.2$ m²

The area may lie outside this range due to the 'large' relative error of length a.

---

1 Expand: (a) $(a + b)^6$

         (b) $(2p - 6q)^5$

2 Evaluate:

(a) $\binom{10}{7}$ (b) $\binom{1000}{997}$

3 In the following expression, find a, b and c. Use Pascal's triangle to explain the equality.

$$\binom{10}{6} = \binom{a}{5} + \binom{9}{b} = \binom{c}{4}$$

4 Find the first four terms of the expansions of:

(a) $(a + b)^{20}$ (b) $(2x - \frac{1}{x^2})^{10}$

5 (a) Use the binomial series to expand $(1 + x)^{-1/2}$ as far as the term in x^3.

  (b) Hence write down the expansion of $(1 - 2x)^{-1/2}$ as far as the term in x^3.

  (c) By taking a suitable value of x, find an approximation for $\frac{1}{\sqrt{0.98}}$ correct to 3 decimal places.

6 Expand the following as far as the term in x^4:

(a) $\frac{1}{(1 + x)^2}$ (b) $\sqrt{(1 - 3x)}$

(c) $\frac{1}{\sqrt{(25 + 100x^2)}}$

giving the values of x for which the expressions are valid.

7 A sheet of paper, area 630 ± 10 cm², has a circular hole, radius 6 ± 0.2 cm, cut from it.

Find the resulting area.

8 Scales, accurate to 50 grams, give a reading of 1.2 kg when a solid metal cube is weighed. The lengths of the sides of the cube are found to be within 1mm of 5 cm. Calculate the density of the cube (in g cm⁻³).

Chain rule

The chain rule is applied in order to differentiate **composite** functions.

If y is a function of u and u is a function of x, then

$$\frac{dy}{dx} = \frac{dy}{du} \times \frac{du}{dx}$$

With practice this technique can be carried out **by inspection** and the answer written down immediately.

■ Differentiate $\sin(x^3 + x)$.

● If $y = \sin(x^3 + x)$ then you can write
$y = \sin u$ where $u = x^3 + x$

$$\frac{dy}{du} = \cos u \text{ and } \frac{du}{dx} = 3x^2 + 1$$

$$\frac{dy}{dx} = \frac{dy}{du} \times \frac{du}{dx} = \cos u \times (3x^2 + 1)$$

$$= (3x^2 + 1) \cos(x^3 + x)$$

The chain rule is a relationship between rates of change and may be applied to problems involving rates of change.

Applications to integration

Being able to differentiate using the chain rule greatly increases the number of functions that can be integrated. Informed guesswork and the knowledge that integration is the reverse of differentiation will help to integrate **some** composite functions.

Useful integrals:

$$\int \sin ax \, dx = -\frac{1}{a} \cos ax + c$$

$$\int \cos ax \, dx = \frac{1}{a} \sin ax + c$$

■ Find $\int_0^1 (1 - 5x)^3 \, dx$

● Try $(1 - 5x)^4$

$$\frac{d}{dx}(1 - 5x)^4 = -20(1 - 5x)^3$$

So $\int_0^1 -20(1 - 5x)^3 \, dx = [(1 - 5x)^4]_0^1$

$$\Rightarrow \int_0^1 (1 - 5x)^3 \, dx = \frac{-1}{20} [(1 - 5x)^4]_0^1$$

$$= \frac{-1}{20} [(-4)^4 - 1^4]$$

$$= -12.75$$

Inverse functions

Application of the chain rule allows differentiation of inverse functions.

$$\frac{dx}{dy} \times \frac{dy}{dx} = 1 \Rightarrow \frac{dx}{dy} = 1 \div \frac{dy}{dx}$$

This result can be used to show that if $y = x^n$ then $\frac{dy}{dx} = nx^{n-1}$ for all n.

Useful derivatives:

$$\frac{d}{dx}(\sin^{-1} x) = \frac{1}{\sqrt{(1 - x^2)}}$$

$$\frac{d}{dx}(\cos^{-1} x) = \frac{-1}{\sqrt{(1 - x^2)}}$$

$$\frac{d}{dx}(\sqrt[n]{x}) = \frac{d}{dx}\left(x^{\frac{1}{n}}\right)$$

$$= \frac{1}{n} x^{\frac{1}{n} - 1}$$

■ Differentiate: (a) $\ln x$

(b) $\sqrt{x}\left(x^2 - \frac{1}{x}\right)$

● (a) $y = \ln x \Rightarrow x = e^y$

$$\frac{dx}{dy} = e^y \Rightarrow \frac{dy}{dx} = \frac{1}{e^y} = \frac{1}{x}$$

(b) $y = x^{1/2}(x^2 - x^{-1}) = x^{5/2} - x^{-1/2}$

$$\frac{dy}{dx} = \frac{5}{2} x^{3/2} - \left(-\frac{1}{2}\right)x^{-3/2}$$

$$= \frac{5}{2}\sqrt{x^3} + \frac{1}{2\sqrt{x^3}}$$

■ A heap of sand is in the shape of a cone with its radius equal to its height. Sand is added to the top of the heap at a rate of 0.5 m³ per hour. Assuming that the heap maintains its shape, find the rate of growth when the heap is 1.5 m tall.

● The volume is given by $V = \frac{1}{3}\pi r^2 h = \frac{1}{3}\pi h^3$

since $r = h$. Also $\frac{dV}{dt} = 0.5$.

You are asked to find $\frac{dh}{dt}$ when $h = 1.5$.

Since $\frac{dV}{dt} = \frac{dV}{dh} \times \frac{dh}{dt}$ and $\frac{dV}{dh} = \frac{1}{3}\pi\,3h^2 = \pi h^2$,

$0.5 = \pi h^2 \times \frac{dh}{dt} \implies \frac{dh}{dt} = \frac{0.5}{\pi h^2}$

When $h = 1.5$, $\frac{dh}{dt} = \frac{0.5}{\pi \times 1.5^2} = 0.07$ m h⁻¹

■ Find $\int \cos^2 x \, dx$.

● $\cos^2 x$ cannot be integrated directly but you can make use of the identity $\cos 2x = 2\cos^2 x - 1$.

Rearranging, $\cos^2 x = \frac{1}{2}(1 + \cos 2x)$

and $\int \cos^2 x \, dx = \frac{1}{2}(x + \frac{1}{2}\sin 2x) + c$

$= \frac{1}{2}x + \frac{1}{4}\sin 2x + c$

■ Find the derivative of $\cos^{-1} x$.

● If $y = \cos^{-1} x$ then $x = \cos y$

$\frac{dx}{dy} = -\sin y \implies \frac{dy}{dx} = \frac{1}{-\sin y}$

$\frac{dy}{dx} = \frac{-1}{\sqrt{(1-\cos^2 y)}} = \frac{-1}{\sqrt{(1-x^2)}}$

---

1 Differentiate: (a) $\sin^2 x$ (b) $\sin x^2$

2 Differentiate the following functions:

(a) $(4x^3 + 3)^7$ (b) $\cos^3 x$ (c) $\frac{1}{(x^4-3)^2}$

(d) $\ln 7x$ (e) $\ln(4x - 1)$ (f) $e^{\sin 2x}$

(g) $\sqrt{(x^2 - 1)}$ (h) $5 \sin^2 3x$

3 Find

(a) $\int (2x + 5)^2 \, dx$ (b) $\int_0^1 e^{3x} \, dx$

(c) $\int \sin 5x \, dx$ (d) $\int_0^\pi \cos \frac{1}{2}x \, dx$

(e) $\int \frac{3}{x} \, dx$ (f) $\int_1^2 x(x^2 + 1) \, dx$

4 A gardener grows leeks which may be considered to be in the shape of a cylinder of radius r and height h, which is ten times the radius. If, when the radius is 2 cm, the leek is growing in height at a rate of 0.7 cm per month, find the rate of increase of its volume.

5 Evaluate the shaded area beneath the graph of $y = 2 \sin 4t$.

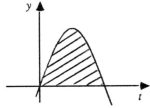

6 A spherical balloon is being inflated so that after t minutes the radius r metres is given by $r = 2 + 0.05t^2$.

(a) Work out expressions for $\frac{dr}{dt}$ and $\frac{dV}{dr}$, where V is the the volume of the balloon.

(b) Combine these to find $\frac{dV}{dt}$ and hence find how fast the volume is increasing after 3 minutes.

7 (a) If $f(x) = \ln(x^2 + a)$ for $a > 0$, find $f'(x)$.

(b) Find the coordinates of any stationary points of $f(x)$.

(c) Use your answers to sketch the graph(s) of $f(x)$.

8

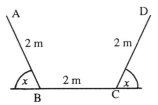

In the diagram, ABCD represents the symmetrical cross-section of a drainage channel ($0 < x < \frac{\pi}{2}$). The sides and the base are made from sheets of metal of width 2 metres, so that AB = BC = CD = 2.

(a) Show that W, the area of the cross-section of the channel, is given by

$W = 4 \sin x + 2 \sin 2x$

(b) Find the value of x for which W is stationary and show that the stationary value is a maximum.

Differential equations

Any equation involving a derivative, such as $\frac{dy}{dx}$ or $\frac{ds}{dt}$, is called a **differential equation.**

The **general solution** of a differential equation will comprise a set of solution curves. If a point is given that lies on the solution curve it enables a **particular solution** to be found.

Differential equations which can be written in the form $\frac{dy}{dx} = f(x)$ may be solved by **inspection** using integration directly.

Direction diagrams

The value of $\frac{dy}{dx}$ is calculated at regularly spaced points in the plane and short line segments drawn to represent the gradients. The resulting diagram gives an impression of the family of solution curves.

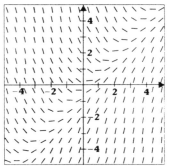

Numerical methods

Approximate solutions to differential equations can be calculated using a **step-by-step** method.

Use the differential equation to calculate the gradient at the starting point. Move in this direction for a fixed step length and calculate the coordinates of the new point. Calculate the gradient at this new point and repeat the process.

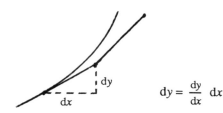

$$dy = \frac{dy}{dx} dx$$

■ For the differential equation $\frac{dy}{dx} = \frac{x}{y}$, find the approximate value of y when $x = 3$ for the particular solution through $(2, 1)$ using the step-by-step method with two steps.

● Using a step length of $dx = 0.5$

x	y	$\frac{dy}{dx}$	dx	dy	$x + dx$	$y + dy$
2	1	2	0.5	1	2.5	2
2.5	2	1.25	0.5	0.625	3	2.625
3	2.625					

So, when $x = 3$, $y \approx 2.63$

Growth and decay

Differential equations of the form $\frac{dy}{dx} = \lambda y$ generate families of solution curves which are graphs of growth functions.

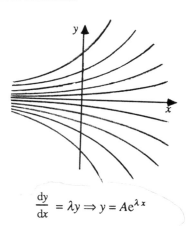

$$\frac{dy}{dx} = \lambda y \Rightarrow y = Ae^{\lambda x}$$

■ Boiling water left for 10 minutes in a room kept at 18°C cools to 60°C. When will it be at 40°C?

● t minutes after boiling, let the water be at y°C above the temperature of the surroundings. Then assuming Newton's law of cooling,

$$\frac{dy}{dt} = \lambda y \Rightarrow y = Ae^{\lambda t}$$

When $t = 0$, $y = 82 \Rightarrow A = 82$
When $t = 10$, $y = 42 \Rightarrow 42 = 82e^{10\lambda}$
$$\Rightarrow 10\lambda = \ln \frac{42}{82}$$
$$\Rightarrow \lambda = -0.067$$

So $y = 82e^{-0.067t}$
When $y = 22$, $22 = 82e^{-0.067t} \Rightarrow t = 19.6$

So the water will have cooled to 40°C after 20 minutes.

■ Solve $x \dfrac{dy}{dx} = 1$ and find the equation of the particular solution curve which passes through the point $(1, 2)$.

● $x \dfrac{dy}{dx} = 1 \Rightarrow \dfrac{dy}{dx} = \dfrac{1}{x}$

and, on integration,
$y = \ln x + c$

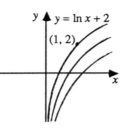

As the curve passes
through $(1, 2)$
$c = 2$ and $y = \ln x + 2$.

■ Sketch the direction diagram for $\dfrac{dy}{dx} = \dfrac{x}{y}$.

● At $(1, 1)$, $\dfrac{dy}{dx} = 1$

At $(1, 2)$, $\dfrac{dy}{dx} = \dfrac{1}{2}$

At $(2, 1)$, $\dfrac{dy}{dx} = 2$ etc.

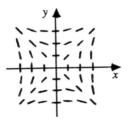

Using a solution sketcher on a computer completes the diagram and indicates the shape of the family of solution curves.

■ A radioactive substance decays at a rate proportional to its mass. When the mass of the sample is 0.020 g, the decay rate is 0.001 g per day.

(a) t days later the mass is m grams. Write down a differential equation and solve it to give m in terms of t for these initial conditions.

(b) Why will the substance never disappear completely? Find its half-life.

● (a) Rate of decay $\dfrac{dm}{dt} = -km \Rightarrow m = Ae^{-kt}$

When $t = 0$, $m = 0.020$ and $\dfrac{dm}{dt} = -0.001$

So $-0.001 = -k \times 0.020 \Rightarrow k = 0.05$
and $0.020 = Ae^0 \Rightarrow A = 0.020$
giving $m = 0.02e^{-0.05t}$

(b)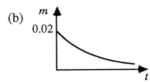

m decays exponentially and, as indicated on the graph, the mass will never entirely disappear.

The length of the substance's half-life is the time it takes to decay to half its original mass, $\dfrac{m}{2}$.

The half-life is given by $0.01 = 0.02e^{-0.05t}$
$$-0.05t = \ln 0.5$$
$$t = 13.8$$
The half-life is nearly 14 days.

---

1 Solve the following differential equations and in each case make a sketch showing some particular solutions.

   (a) $\dfrac{dy}{dx} = 3x^2$ (b) $\dfrac{dy}{dx} = e^{2x}$ (c) $\dfrac{dy}{dx} = \sin 3x$

2 For each of the differential equations given below find the particular solution which passes through the point $(1, 2)$.

   (a) $x \dfrac{dy}{dx} = 2x^2 + 1$ (b) $\dfrac{dy}{dx} = \cos (2x - 2)$

3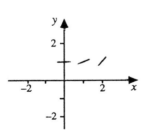

   (a) Complete this direction diagram for $\dfrac{dy}{dx} = \dfrac{1}{2}xy$.

   (b) Sketch a solution curve through $(0, 1)$.

4 Using a step-by-step method, continue this table of values to generate a numerical solution to the equation $\dfrac{dy}{dx} = 3x - y$ and find the approximate value of y when $x = 2$.

x	y	$\dfrac{dy}{dx}$	dx	dy	$x + dx$	$y + dy$
1	3.5	–0.5	0.2	–0.1	1.2	3.4
1.2						

5 40 litres of water flows out of a rainwater tank in the first 4 minutes. The tank initially held 250 litres and the rate of flow is believed to be proportional to the volume of water remaining in the tank.

   Describe the relationship in symbols, stating their units and meaning. Find an expression for the volume of water remaining after t minutes.

6 $\dfrac{dy}{dx} = y + 1$ and $y = 2$ when $x = 1$. Use a numerical method, with steps of 0.2, to estimate the value of y when $x = 2$.

CALCULUS METHODS

Parametric equations

An (x, y) graph can be drawn when x and y are described in terms of a third variable.

The third variable is called a **parameter** and the equations defining x and y in terms of the parameter are **parametric equations.** It is often possible to obtain the Cartesian equation of the curve from the parametric equations.

■ The parametric equations of a curve are given in terms of a parameter, t, as: $x = 2t$, $y = t^2$. Work out the Cartesian equation of the curve.

● $x = 2t \Rightarrow t = \frac{1}{2} x$

So $y = (\frac{1}{2} x)^2 = \frac{1}{4}x^2$

Circles and ellipses

	Cartesian equation	Parametric equations
Circle centre $(0, 0)$ and radius r	$x^2 + y^2 = r^2$	$x = r \cos \theta, y = r \sin \theta$
Ellipse: 	$\dfrac{x^2}{a^2} + \dfrac{y^2}{b^2} = 1$	$x = a \cos \theta, y = b \sin \theta$

Trigonometric identities

Trigonometric identities can be used to convert some parametric equations (including those for circles and ellipses) into Cartesian equations. The identities which may be useful are:

$$\cos^2 \theta + \sin^2 \theta = 1 \qquad 1 + \tan^2 \theta = \sec^2 \theta \qquad \cot^2 \theta + 1 = \operatorname{cosec}^2 \theta$$

where: $\sec \theta = \dfrac{1}{\cos \theta}$ $\qquad \operatorname{cosec} \theta = \dfrac{1}{\sin \theta}$ $\qquad \cot \theta = \dfrac{1}{\tan \theta}$

Parametric differentiation

When x and y are given in terms of a parameter, it is not necessary to work out the Cartesian equation in order to find $\dfrac{dy}{dx}$. A version of the chain rule can be used.

If x and y are functions of t:

$$\frac{dy}{dx} = \frac{dy}{dt} \times \frac{dt}{dx} \quad \text{or} \quad \frac{dy}{dx} = \frac{dy}{dt} \div \frac{dx}{dt}$$

Velocity vectors

If a particle is moving such that its position vector is given by: $r = \begin{bmatrix} x \\ y \end{bmatrix}$

and x and y are functions of time, t, then the velocity vector of the object at time t is

$v = \begin{bmatrix} \dot{x} \\ \dot{y} \end{bmatrix}$, where $\dot{x} = \dfrac{dx}{dt}$ and $\dot{y} = \dfrac{dy}{dt}$.

The gradient of the velocity vector is $\dfrac{dy}{dx} = \dfrac{\dot{y}}{\dot{x}}$.

■ A curve has parametric equations:

$$x = \sec \theta - 1, \quad y = \tan^2 \theta$$

(a) Calculate the values of x and y for $\theta = 0°$, $15°$, $30°$, $45°$ and $60°$ and so sketch the part of the curve for which $0° \le \theta \le 60°$.

(b) Show that the Cartesian equation of the curve is $y + 1 = (x + 1)^2$.

● (a)

θ	0	15°	30°	45°	60°
x	0	0.035	0.15	0.41	1
y	0	0.072	0.33	1	3

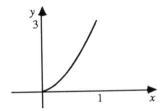

(b) $\sec \theta = x + 1$, $\tan^2 \theta = y$

So, since $1 + \tan^2 \theta = \sec^2 \theta$
$$y + 1 = (x + 1)^2$$

■ Work out an expression for $\dfrac{dy}{dx}$ for the curve defined by:

$$x = \sin 2t, \quad y = \sin^2 t$$

Hence find the gradient of the curve when $t = \dfrac{1}{8}\pi$.

● $\dfrac{dy}{dt} = 2 \sin t \cos t$

$\dfrac{dx}{dt} = 2 \cos 2t$

$\dfrac{dy}{dt} = \dfrac{dy}{dt} \div \dfrac{dx}{dt} = \dfrac{\sin t \cos t}{\cos 2t}$

At $t = \dfrac{\pi}{8}$, the gradient is:

$$\dfrac{\sin(\pi/8)\cos(\pi/8)}{\cos(\pi/4)} = 0.5$$

■ The position vector of a particle after t seconds is:

$$\begin{bmatrix} 20t \\ 12 - t^2 \end{bmatrix} \text{ metres.}$$

Find the velocity and speed after 2 seconds.

● The velocity vector at time t is $\begin{bmatrix} 20 \\ -2t \end{bmatrix}$ m s⁻¹.

After 2 seconds the velocity is $\begin{bmatrix} 20 \\ -4 \end{bmatrix}$ m s⁻¹.

The speed is $\sqrt{(20^2 + 4^2)} = 20.4$ m s⁻¹ (to 1 d.p.).

---

1 (a) Plot the curve given by the parametric equations

$$x = 3 - t^2, y = -2t \text{ for } 0 \le t \le 4$$

(b) What is the Cartesian equation of the curve?

2 An ellipse is described by the parametric equations:

$$x = 3 \cos \theta, y = 5 \sin \theta$$

Write down the lengths of the major and minor axes and sketch the ellipse.

3 An ellipse has parametric equations:

$$x = 3 \cos \theta, \ y = 2 \sin \theta$$

(a) Find $\dfrac{dy}{dx}$ in terms of θ.

(b) Show that the equation of the tangent at the point with parameter θ is:

$$2x \cos \theta + 3y \sin \theta = 6$$

(c) What is the equation of the tangent at the point $(3, 0)$?

4 Work out the Cartesian equations of the curves whose parametric equations are:

(a) $x = \dfrac{1}{2}t, \ y = 3t^2 - 5t$

(b) $x = 5t, \ y = 6 - \dfrac{10}{t}$

5 Find $\dfrac{dy}{dx}$ in terms of t if:

$$x = \ln 2t, y = 2t + 3 \ln t$$

and work out the equation of the tangent to the curve at $t = 1$.

6 The path of a skater is described by the parametric equations:

$$x = 3t, y = \dfrac{t^3}{10}$$

where t is the time in seconds and distances are measured in metres.

(a) Where is the skater when $t = 10$?

(b) How far is the skater from her starting point when $t = 3$?

(c) What is the velocity of the skater when $t = 2$?

(d) Calculate the speed of the skater when $t = 1$.

Product rule and quotient rule

If u and v are functions of x,

when $y = uv$, $\frac{dy}{dx}$ can be found by using the **product rule**:

$$\frac{dy}{dx} = v \frac{du}{dx} + u \frac{dv}{dx}$$

when $y = \frac{u}{v}$, $\frac{dy}{dx}$ can be found by using the **quotient rule**:

$$\frac{dy}{dx} = \frac{v \frac{du}{dx} - u \frac{dv}{dx}}{v^2}$$

Product rule or chain rule

It is important to check carefully to ensure that the correct choice is made between using the chain rule or the product rule.

The **chain rule** differentiates the function of a function.

The **product rule** differentiates the product of two functions.

■ Which rule must you use to differentiate:

(i) $x^2 e^x$ (ii) e^{x^2} ?

● (i) $x^2 e^x$ can be written as uv, with $u = x^2$; $v = e^x$. The product rule is needed.

(ii) e^{x^2} is $fg(x)$, where $g(x) = x^2$ and $f(x) = e^x$. The chain rule is needed.

Product rule and chain rule

It is sometimes necessary to use both the chain rule and the product rule in order to find a derivative.

For example, $y = \sin^2 x \cos x$ can be written as $y = uv$ with $u = \sin^2 x$ and $v = \cos x$. The chain rule must then be used to find $\frac{du}{dx}$.

Implicit differentiation

If $y = f(x)$, y is stated **explicitly** and $\frac{dy}{dx}$ can be found by differentiation.

$\frac{dy}{dx}$ can also be found when y is stated **implicitly**.

For example:

$$x^3 + 5x + y^2 - 4y = 6 \quad ①$$

By using the chain rule, both sides of ① can be differentiated with respect to x: $\Big\}$

$$3x^2 + 5 + 2y \frac{dy}{dx} - 4 \frac{dy}{dx} = 0$$

$3x^2 + 5 + 2y \frac{dy}{dx} - 4 \frac{dy}{dx} = 0$ can be rearranged to give $\frac{dy}{dx} = \frac{3x^2 + 5}{4 - 2y}$.

If the implicit function contains terms in xy, xy^2, etc., the product rule and the chain rule must be used.

$$x^2 + \overbrace{4xy} - y^2 + 6x = 10 \quad \text{leads to:}$$

$$2x + 4y + 4x \frac{dy}{dx} - 2y \frac{dy}{dx} + 6 = 0$$

■ Find $\dfrac{dy}{dx}$ if:

(a) $y = \dfrac{x+5}{x^2-3}$

(b) $y = (3x-2)^2 (2x+1)^3$

● (a) $y = \dfrac{u}{v}$, with:

$u = x+5; \quad \dfrac{du}{dx} = 1$

$v = x^2-3; \quad \dfrac{dv}{dx} = 2x$

$\dfrac{dy}{dx} = \dfrac{1(x^2-3)-2x(x+5)}{(x^2-3)^2}$

$= \dfrac{-x^2-10x-3}{(x^2-3)^2}$

(b) $y = uv$, with:

$u = (3x-2)^2, \quad \dfrac{du}{dx} = 6(3x-2)$

$v = (2x+1)^3, \quad \dfrac{dv}{dx} = 6(2x+1)^2$

$\dfrac{dy}{dx} = 6(2x+1)^3 (3x-2) + 6(3x-2)^2 (2x+1)^2$

$= 6(2x+1)^2 (3x-2) (2x+1+3x-2)$

$= 6(2x+1)^2 (3x-2) (5x-1)$

■ Find the coordinates of the minimum point of the graph of $y = (x-3)\,e^x$.

● Put $u = x-3$, $v = e^x$

$\dfrac{du}{dx} = 1, \quad \dfrac{dv}{dx} = e^x$

$\dfrac{dy}{dx} = e^x + (x-3)e^x$

$= e^x (1+x-3) = e^x (x-2)$

$e^x (x-2) = 0 \Rightarrow x = 2 \quad (e^x \neq 0)$

$x = 1.9 \Rightarrow \dfrac{dy}{dx} = -0.67;$

$x = 2.1 \Rightarrow \dfrac{dy}{dx} = 0.82$

So $x = 2$ gives a minimum point, $x = 2 \Rightarrow y = -7.39$ (to 3 s.f.), so the coordinates of the minimum point are $(2, -7.39)$.

■ A curve is defined by the equation:

$x^2 + xy - 2y^2 = 13$

Work out the gradient of the curve at the point (5, 4).

● $2x + y + x\,\dfrac{dy}{dx} - 4y\,\dfrac{dy}{dx} = 0$

$\dfrac{dy}{dx} = \dfrac{2x+y}{4y-x}$

The gradient at (5, 4) is $\dfrac{14}{11} = 1.27$ (to 3 s.f.).

---

1 Find $\dfrac{dy}{dx}$ when:

(a) $y = \ln(\sin x)$

(b) $y = \cos^3 x$

(c) $y = \cos x^3$

(d) $y = \dfrac{1}{(4x-3)^2}$

(e) $4x^2 + 3x^2 y - y^2 = 8$

(f) $x = 3t-5, \; y = 4t^2$

(g) $y = x^2 e^{3x}$

(h) $y = \dfrac{2x-3}{4x+5}$

(i) $y = \sec x = \dfrac{1}{\cos x}$

(j) $4x^2 - 5x - 2y^2 + 8y = 23$

(k) $y = 2 \sin 2x \cos 3x$

(l) $x = 2\cos\theta, \; y = 4\sin 2\theta$

2 Use the quotient rule to show the derivative of $\tan x$ is $\sec^2 x$.

3 The graph of $y = e^{0.5x} \sin x$ has a local maximum and a local minimum in the region $0 \le x \le 6$.

Find the coordinates of these points, showing clearly (without the aid of a graph) which is which.

4 Find the positive value of x for which

$\dfrac{x}{(3x+2)^3}$

takes its greatest value. (You need not prove it is a maximum rather than a minimum.)

5 (a) Find $\dfrac{dy}{dx}$ for the curve:

$x^3 - y^3 + 3x + 2y + 7 = 0$

(b) Work out the equation of the tangent to the curve at the point (2, 3).

6 $f(x) = \dfrac{\sin x + 1}{\cos x + 3}$

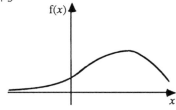

(a) Work out $f'(x)$.

(b) Work out the equation of the tangent to $y = f(x)$ at $x = 0$.

Thin slabs

The volume of a solid can be found by thinking of the solid as the sum of a large number of thin slabs.

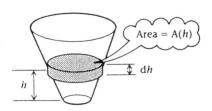

$$V = \int A(h)\, dh$$

Solids of revolution

Solids of revolution are formed by rotating areas about the x-axis or the y-axis.

If the area between $y = f(x)$ and the x-axis for $a \le x \le b$ is rotated about the x-axis, the solid has volume:

$$V = \pi \int_a^b y^2\, dx$$

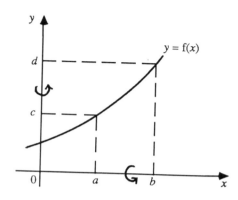

If the area betwen $y = f(x)$ and the y-axis for $c \le y \le d$ is roatated about the y-axis, the solid has volume:

$$V = \pi \int_c^d x^2\, dy$$

Integration by inspection

It is sometimes possible to integrate functions by **inspection.**

This involves thinking of the function which can be differentiated to give a multiple of the function you wish to integrate.

■ Integrate $xe^{(x^2-3)}$

● $\dfrac{d}{dx}(e^{(x^2-3)}) = 2xe^{(x^2-3)}$

So $\int 2xe^{(x^2-3)}\, dx = e^{(x^2-3)} + c$

and $\int xe^{(x^2-3)}\, dx = 0.5e^{(x^2-3)} + c$

Trigonometric identities

Trigonometric identities can sometimes be used to convert products of functions into sums or differences of functions and so make integration possible. These identities are:

$$2 \cos A \cos B = \cos (A + B) + \cos (A - B)$$

$$2 \sin A \sin B = - \cos (A + B) + \cos (A - B)$$

$$2 \sin A \cos B = \sin (A + B) + \sin (A - B)$$

The double angle formulas can be arranged to make it possible to integrate $\cos^2 x$, $\sin^2 x$ and $\sin x \cos x$.

$$\cos 2x = 2 \cos^2 x - 1 \Rightarrow \cos^2 x = \frac{1}{2}(1 + \cos 2x)$$

$$\cos 2x = 1 - 2 \sin^2 x \Rightarrow \sin^2 x = \frac{1}{2}(1 - \cos 2x)$$

$$\sin 2x = 2 \sin x \cos x \Rightarrow \sin x \cos x = \frac{1}{2}\sin 2x$$

■ A square pyramid of height H is such that the length, x cm, of a side of its cross-section is twice the height of the section above the apex.

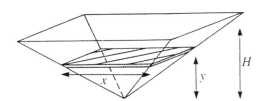

Given that the volume of the shaded slab is x^2 dy, show that the volume of the whole pyramid is $\frac{4}{3}H^3$.

● $V = \int_0^H x^2 \, dy$

But $x = 2y$, so $x^2 = 4y^2$

So $V = \int_0^H 4y^2 \, dy$

$= 4 \left[\frac{y^3}{3} \right]_0^H$

$= \frac{4}{3} H^3$

■ (a) Evaluate $\int_0^{\pi/2} \sin x \cos x \, dx$

    (i) by using the formula for $\sin 2x$;

    (ii) by inspection.

● (a) (i) $\sin x \cos x = \frac{1}{2} \sin 2x$

$\int_0^{\pi/2} \frac{1}{2} \sin 2x \, dx = \left[-\frac{1}{4} \cos 2x \right]_0^{\pi/2} = 0.5$

    (ii) $\frac{d}{dx} (\sin^2 x) = 2 \sin x \cos x$

$\int_0^{\pi/2} \sin x \cos x \, dx = \left[\frac{1}{2} \sin^2 x \right]_0^{\pi/2} = 0.5$

■ A glass bowl is in the shape of the curve $y = \sqrt{x}$ for $0 \le x \le 9$, rotated $360°$ about the y-axis.

Calculate the volume of water needed to fill it.

● $V = \pi \int_{x=0}^{x=9} x^2 \, dy$

$y = \sqrt{x} \Rightarrow x^2 = y^4$
$0 \le x \le 9 \Rightarrow 0 \le y \le 3$

So $V = \pi \int_0^3 y^4 \, dy$

$= \left[\pi \frac{y^5}{5} \right]_0^3 \approx 153$ units

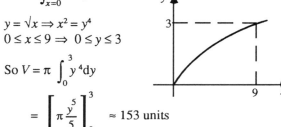

---

1 (a) Find $\frac{dy}{dx}$ for:

    (i) $y = \sin(x^3)$ (ii) $y = \sin^3 x$

  (b) Use your answers to find:

$\int x^2 \cos(x^3) \, dx$ and $\int \sin^2 x \cos x \, dx$

2 (a) Write down $\int \cos 2\theta \, d\theta$.

  (b) Use the identity:

$\cos 2\theta = 1 - 2\sin^2 \theta$

to find $\int \sin^2 \theta \, d\theta$.

  (c) Evaluate $\int_0^{\pi/2} 0.5 \cos^2 \theta \, d\theta$.

3 (a) Write $\sin 2x \cos 4x$ as the sum of two sines and hence evaluate:

$\int_{0.5}^1 \sin 2x \cos 4x \, dx$

  (b) Evaluate $\int_1^{1.5} \cos x \cos 3x \, dx$.

4 Work out the volume obtained by rotating the graph of $y = \sin x$ between $x = 0$ and $x = \frac{\pi}{2}$ about the x-axis.

5 Integrate by inspection:

  (a) $x^2 e^{x^3}$ (b) $(7x + 12)^{12}$

  (c) e^{-5x} (d) $(2x + 3)(x^2 + 3x)^4$

  (e) $\frac{2x+3}{x^2 + 3x}$ (f) $\sin 6x$

  (g) $x \cos(3x^2 + 4)$

6

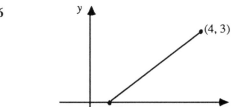

The line $y = x - 1$ between $(1, 0)$ and $(4, 3)$ is rotated by $360°$ about the y-axis to form a solid. Find the volume of this solid.

Integration by parts

The formula for the product rule for differentation can be rearranged into a form which is helpful when integrating some products of functions.

$$\int u \frac{dv}{dx} \, dx = uv - \int v \frac{du}{dx} \, dx$$

When applying the by parts formula it is usually important to make sure that 'u' becomes simpler when differentiated and that '$\frac{dv}{dx}$' has a straightforward integral. For example, to integrate the product $x \cos 2x$, u would be set equal to x and $\frac{dv}{dx}$ would be set equal to $\cos 2x$.

Integration by substitution

The formula for the chain rule for differentiation can be used to establish the formula

$$\int f(x) \, dx = \int f(x) \frac{dx}{du} \, du$$

The expression $f(x) \frac{dx}{du}$ must be expressed entirely in terms of the new variable u. A suitable choice for the function u is often apparent because the expression for $\frac{du}{dx}$ can be seen in the original integral. For example, to find

$$\int x^2 \sqrt{(x^3 + 1)} \, dx,$$

the substitution $u = x^3 + 1$ is suggested by the fact that $\frac{du}{dx} = 3x^2$. Then

$$\int x^2 \sqrt{(x^3 + 1)} \, dx = \int \sqrt{u} \, \frac{1}{3} \, du = \frac{1}{3} \times \frac{2}{3} u^{\frac{3}{2}} + c = \frac{2}{9} (x^3 + 1)^{\frac{3}{2}} + c$$

The reciprocal function

If the graph of $f(x)$ is continuous between a and b, the method of substitution can be used to show that

$$\int_a^b \frac{f'(x)}{f(x)} \, dx = \left[\ln | f(x) | \right]_a^b$$

■ Find $\displaystyle\int_1^2 \frac{3}{x-4} \, dx$

● $\displaystyle\int_1^2 \frac{3}{x-4} \, dx = 3 \left[\ln | x - 4 | \right]_1^2$

$\qquad\qquad\qquad = 3 \, (\ln 2 - \ln 3)$

$\qquad\qquad\qquad = 3 \ln \frac{2}{3}$

Partial fractions

Splitting a polynomial fraction into partial fractions enables you to integrate the polynomial fraction easily using the fact that:

$$\int \frac{1}{ax+b} \, dx = \frac{1}{a} \ln | ax + b | + c$$

See 3 methods in notes

To split a polynomial fraction such as $\dfrac{2x^2 + 4x + 3}{x^2 + x - 2}$ into partial fractions:

(1) Divide out until the degree of the numerator is less than the degree of the denominator.

$$\frac{2x^2 + 4x + 3}{x^2 + x - 2} = \frac{2(x^2 + x - 2) + 2x + 7}{x^2 + x - 2} = 2 + \frac{2x+7}{x^2 + x - 2}$$

(2) Factorise the denominator and solve equations for the constants.

$$\frac{A}{x+2} + \frac{B}{x-1} = \frac{2x+7}{(x+2)(x-1)}$$

$\Rightarrow A \, (x - 1) + B \, (x + 2) = 2x + 7$
$\Rightarrow A = -1, \; B = 3$

So $\dfrac{2x^2 + 4x + 3}{x^2 + x - 2} = 2 - \dfrac{1}{x+2} + \dfrac{3}{x-1}$

■ Find: (a) $\int \sin^n x \cos x \, dx$

    (b) $\int x \ln x \, dx$

● (a) Let $u = \sin x$, $\dfrac{du}{dx} = \cos x$

    $I = \int u^n \, du = \dfrac{1}{n+1} u^{n+1} + c$

    $= \dfrac{1}{n+1} \sin^{n+1} x + c$

  (b) Let $u = \ln x$, $\dfrac{dv}{dx} = x$

    Then $\dfrac{du}{dx} = \dfrac{1}{x}$, $v = \dfrac{1}{2} x^2$

    $I = \dfrac{1}{2} x^2 \ln x - \int \dfrac{1}{2} x^2 \, \dfrac{1}{x} \, dx$

    $= \dfrac{1}{2} x^2 \ln x - \dfrac{1}{2} \int x \, dx$

    $= \dfrac{1}{2} x^2 \ln x - \dfrac{1}{4} x^2 + c$

■ Rewrite $\dfrac{x^2 + 7x + 5}{x^2 + 5x + 6}$ in the form $A + \dfrac{px + q}{x^2 + 5x + 6}$

and then in the form $A + \dfrac{B}{x+2} + \dfrac{C}{x+3}$.

Hence find $\int \dfrac{x^2 + 7x + 5}{x^2 + 5x + 6} \, dx$.

● $\dfrac{x^2 + 7x + 5}{x^2 + 5x + 6} = 1 + \dfrac{2x - 1}{x^2 + 5x + 6}$

  $\dfrac{2x - 1}{x^2 + 5x + 6} = \dfrac{B}{x+2} + \dfrac{C}{x+3}$

  $\Leftrightarrow 2x - 1 = B(x + 3) + C(x + 2)$

  Let $x = -2$, then $-5 = B$. Let $x = -3$, then $-7 = -C$.

  Then $\dfrac{x^2 + 7x + 5}{x^2 + 5x + 6} = 1 - \dfrac{5}{x+2} + \dfrac{7}{x+3}$

  $\int \dfrac{x^2 + 7x + 5}{x^2 + 5x + 6} \, dx = x - 5 \ln |x + 2| + 7 \ln |x + 3| + c$

---

1 Integrate:

  (a) $\sin 5x$ (b) $x \sin 5x$

  (c) $x^4 (3 - x)$ (d) $\dfrac{x^2}{x^3 - 3}$

  (e) $x e^{x^2}$ (f) $\dfrac{1}{5x + 2}$

  (g) $x e^{3x}$ (h) $\sec^2(4x - 3)$

  (i) $\cos^2 3x$ (j) $\dfrac{3}{2 - x}$

  (k) $2\sqrt{x} - \dfrac{3}{\sqrt{x}}$ (l) $(5x - 3)^{10}$

  (m) $\dfrac{x}{\sqrt{(x^2 + 4)}}$ (n) $\dfrac{1}{\sqrt{(2x + 4)}}$

  (o) e^{-8x} (p) $\dfrac{x^3 - 3x}{x^2}$

  (q) $x(3 + 2x)^4$ (r) $(5 - \cos 3x)^3 \sin 3x$

  (s) $x^3 \ln x$ (t) $x^2 \sin x$

2 Integrate:

  (a) $\dfrac{e^{2x}}{3}$ (b) $\dfrac{2}{e^{3x}}$

3 By parts, or otherwise, evaluate $\int_0^1 x (x + 3)^5 \, dx$.

4 Verify that the curves $y = \dfrac{1}{x^2}$ and $y = \dfrac{1}{4x - 3}$ intersect at $x = 1$ and $x = 3$ and find the area enclosed between them in this interval.

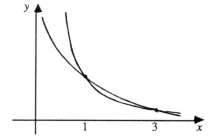

5 Differentiate $\sin (3x^2 + 4)$ and hence, or otherwise, evaluate $\int_0^6 x \cos (3x^2 + 4) \, dx$.

6 By using the substitution $x = 4 \sin u$, or otherwise, find $\int \dfrac{1}{\sqrt{(16 - x^2)}} \, dx$.

7 (a) Express $\dfrac{5x^2 - 59}{x^2 - x - 12}$ in the form $A + \dfrac{px + q}{x^2 - x - 12}$

    and then in the form:

    $A + \dfrac{B}{x+b} + \dfrac{C}{x+c}$

  (b) Hence find $\int_0^1 \dfrac{5x^2 - 59}{x^2 - x - 12} \, dx$.

Taylor's first approximation

The first approximation to a function at a point is the linear function which passes through the point and has the same gradient.

If $f(x) \approx a + bx$ at $x = p$, then $b = f'(p)$ and $a + bp = f(p)$.

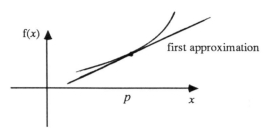

first approximation

Newton-Raphson

This iterative method is used to find a solution of $f(x) = 0$.

(a) Locate the roots by sketching a graph.

(b) Choose x_1 close to the solution.

(c) Find $x_2, x_3, \dots$ using the iterative formula

$$x_{n+1} = x_n - \frac{f(x_n)}{f'(x_n)}$$

■ Find (to 6 d.p.) the positive solution of
$$e^x = 5 \cos x$$

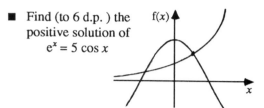

● Let $f(x) = e^x - 5 \cos x$
$\Rightarrow f'(x) = e^x + 5 \sin x$
$x_1 = 1$
$x_2 = 0.9975785$
$x_3 = 0.9975762$
$x_4 = 0.9975762$
$\Rightarrow \quad x = 0.997576$ (to 6 d.p.)

Maclaurin's series

If $f(x)$ can be differentiated n times at $x = 0$, then the approximation

$$f(x) \approx f(0) + f'(0)x + \frac{f''(0)}{2!} x^2 + \dots + \frac{f^{(n)}(0)}{n!} x^n$$

will be good for values of x close to $x = 0$. $f^{(n)}(x)$ is the nth differential of $f(x)$; for instance, $f^{(4)}(x)$ is the 4th differential of $f(x)$.

Particular approximations:

$$\sin x = x - \frac{x^3}{3!} + \frac{x^5}{5!} - \frac{x^7}{7!} + \dots$$

$$\cos x = 1 - \frac{x^2}{2!} + \frac{x^4}{4!} - \frac{x^6}{6!} + \dots$$

$$e^x = 1 + x + \frac{x^2}{2!} + \frac{x^3}{3!} + \frac{x^4}{4!} + \dots$$

$$\ln(1 + x) = x - \frac{x^2}{2} + \frac{x^3}{3} - \frac{x^4}{4} + \dots \quad (-1 < x < 1)$$

The binomial series is valid when $-1 < x < 1$:

$$(1 + x)^n = 1 + nx + \frac{n(n-1)}{2!} x^2 + \frac{n(n-1)(n-2)}{3!} x^3 + \dots$$

First principles

A general formula for $f'(x)$ can be found from first principles using the definition

$$f'(x) = \lim_{h \to 0} \frac{f(x+h) - f(x)}{h}$$

■ For $y = x^3$, find $\frac{dy}{dx}$ from first principles.

● $\frac{dy}{dx} = \lim_{h \to 0} \dfrac{(x^3 + 3hx^2 + 3h^2x + h^3) - x^3}{h}$

$\quad = \lim_{h \to 0} (3x^2 + 3hx + h^2) = 3x^2$

■ For the function $y = e^{x/2} - \cos x$ at $x = 1$

(a) obtain Taylor's first approximation;

(b) find the percentage error at $x = 1.2$.

● (a) $f'(x) = \frac{1}{2}e^{x/2} + \sin x$

$f'(1) = 1.6658, f(1) = 1.1084$

$\frac{y - 1.1084}{x - 1} = 1.6658 \Rightarrow y = 1.6658x - 0.5574$

(b) $\frac{(e^{0.6} - \cos 1.2) - (1.6658 \times 1.2 - 0.5574)}{(e^{0.6} - \cos 1.2)} \times 100$

$= 1.25\%$

■ Use the first terms of a Maclaurin series

to obtain an approximate value for $\int_0^1 \cos \sqrt{x}\, dx$.

● $\cos x = 1 - \frac{x^2}{2!} + \frac{x^4}{4!} - \frac{x^6}{6!} + \ldots$

$\Rightarrow \cos \sqrt{x} = 1 - \frac{x}{2} + \frac{x^2}{24} - \frac{x^3}{720}$

$\int_0^1 \cos \sqrt{x}\, dx \approx \int_0^1 \left(1 - \frac{x}{2} + \frac{x^2}{24} - \frac{x^3}{720}\right) dx$

$\approx \left[x - \frac{x^2}{4} + \frac{x^3}{72} - \frac{x^4}{2880}\right]_0^1$

$= 0.764$

---

1 Use the Newton-Raphson method, with $x_0 = 1$, to find the positive solution to the equation

$$\sin 2x = \frac{x^2}{3}$$

to 3 decimal places, showing the value of your approximation after each application of the method.

2 What is the Taylor's first approximation to the function

$$y = e^{2x} \quad \text{when } x = 1?$$

3 Derive the Maclaurin's series for $f(x) = \cos 3x$ up to the term involving x^4.

4 A circular disc, centre O, is to be divided by a straight cut AB so that the smaller area ACB is $\frac{1}{10}$ of the area of the whole circle.

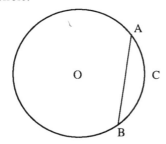

(a) Show that, if angle AOB $= \theta$ radians, then

$$\theta - \frac{1}{5}\pi = \sin \theta$$

(b) Draw sketch graphs to show that this equation has a solution near $\theta = \frac{1}{2}\pi$.

(c) Use the Newton-Raphson method once, with $\frac{1}{2}\pi$ as the first approximation, to obtain a better approximation of θ.

[SMP 1988]

5 For the function $f(x) = \tan x$:

(a) find the first two non zero terms of the Maclaurin's series by differentiation and evaluation of $f(0), f'(0), \ldots$;

(b) find the percentage error involved using this approximation when $x = 0.5$.

6 Use the Taylor approximation

$$e^{-h} \approx 1 - h$$

to show that, when h is a small positive number,

$$\sqrt{(1 - e^{-h})} \approx \sqrt{h}$$

(a) Use a calculator to compare the values of the two expressions when $h = 0.1$.

(b) A function f is defined by $f(x) = \sqrt{(1 - e^{-x})}$, and its domain is the set of values of x for which this expression is defined. Find

(i) the domain of f;

(ii) the range of f.

[Your answer should make clear whether or not the numbers at the ends of the intervals are included.]

(c) Sketch the graph of $y = f(x)$.

[SMP 1986]

MISCELLANEOUS EXERCISES

FOUNDATIONS

1   The graph shown has equation
$$y = (x + p)^2 + q$$

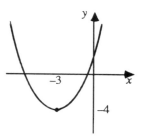

(a)   Find p and q.

(b)   The equation can also be written as $y = x^2 + bx + c$. Find b and c.

(c)   The equation can also be written as $y = (x + k)(x + h)$. Find k and h.

(d)   Where does the graph cut the x-axis?

(e)   Where does the graph cut the y-axis?

2   $g(x) = x^3 + 6x^2 + 11x + 6$

(a)   Find $g(-3)$.

(b)   Factorise $g(x)$.

(c)   Find $g(x - 2)$ in the form $x^3 + bx^2 + cx + d$.

(d)   How are the graphs of $y = g(x)$ and $y = g(x - 2)$ related?

3   The cost of running a train from London to Brighton at an average speed of V miles per hour can be computed as the sum of three components:

- fixed costs (e.g. track maintenance, signalling);
- wages (approximately proportional to the time taken for the journey);
- fuel (approximately proportional to the square of the average speed).

(a)   Write a formula for the total cost in terms of V and appropriate constants.

(b)   Sketch three separate graphs (which must be clearly labelled) showing how each of the three components varies with V. Then sketch a fourth graph showing how the total cost varies with V.

[SMP 1982]

4   (a)   Find the first term in the geometric series 1, 4, 16, ... which is greater than 3000.

(b)   Find the first triangular number which is greater than 3000.

5   Rajish deposits £200 annually in an account which earns 11.5% compound interest per annum. How much money will he have in his account immediately after making the 12th deposit?

6   Claire invests £2000 annually at 10.5% compound interest per annum.

(a)   What is the total amount of her money at the end of 8 years?

(b)   What is the total amount of money at the end of n years?

(c)   How many years will it take for the total sum of money in the account to be three times the total amount of investment?

7   Two positive numbers differ by 3 and have a product of 20. Let x be the larger of the numbers.

(a)   Write down an equation in terms of x.

(b)   Use your answer to part(a) to find the 2 numbers.

8   An explorer sets out for a destination 35 miles away through the jungle. He walks 10 miles in the first day, but on each succeeding day he can walk only 70% of the distance he walked the day before. Find whether he will ever reach his destination, giving a reason for your answer.

[SMP 1979]

9   The recurring decimal $0.5\dot{1}\dot{3}$ signifies the sum to infinity of the series

$$0.513 + 0.000\ 513 + 0.000\ 000\ 513 + \dots$$

Find, in its simplest form, the fraction which gives this recurring decimal.

[SMP 1981]

10   A college student has a pint of milk delivered to her flat each day during the term. When she gets back in the evening she drinks three-quarters of all the milk she has and leaves the rest overnight. She never drinks milk at any other time.

(a)   If u_n denotes the amount of milk she has left at the end of the nth day of term, give (in fraction form) the values of u_1, u_2 and u_3.

(b)   Write an equation connecting u_n with u_{n-1} and show that it is satisfied by the formula

$$u_n = \frac{1}{3} - \frac{1}{3}\left(\frac{1}{4}\right)^n$$

[SMP 1986]

INTRODUCTORY CALCULUS

1   A cattle trough is 3 m long and made of concrete, cast as shown in the diagram.

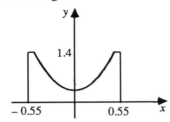

The equation of the interior of the trough is $y = 0.4 + 4x^2$. Find the volume of concrete in the trough.

2   The diagram illustrates the path of a golf ball struck with speed u from a point O on level ground.

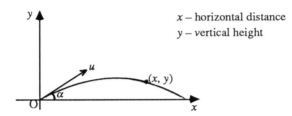

x – horizontal distance
y – vertical height

The equation of the path can be shown to be

$$y = x \tan \alpha - \frac{5x^2}{u^2 \cos^2 \alpha}$$

(a)   The initial velocity, u, of the ball is 30 ms⁻¹ and the angle of projection, α, is 45°. Show that the equation above becomes

$$y = x - \frac{x^2}{90}$$

(b)   Find the maximum height the golf ball will reach.

(c)   Show that the ball will just clear a tree of height 20 m if the tree is 30 m from O.

(d)   Find how far from O the ball lands.

(e)   Draw a sketch of the flight of the ball, clearly labelling the information found in (b), (c) and (d).

3   A sketch graph of a function f is as shown.

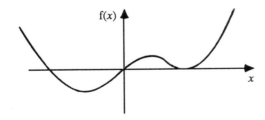

Sketch the graph of the gradient function of f.

4   (a)   Find the gradient function for

$$y = 3x^4 - 4x^3$$

(b)   Hence find the coordinates of the stationary points on the curve $y = 3x^4 - 4x^3$.

(c)   Sketch the curve.

5   Find the equation of the tangent at (2, 8) to the curve

$$y = 3x^2 - x - 2$$

6   A flood protection scheme on an otherwise flat coast consists of a long mound of earth with a ditch scooped out on the landward side. The cross-section of the mound and the ditch can be approximated by a cubic graph with equation $y = x^3 - 4x$, the units being metres.

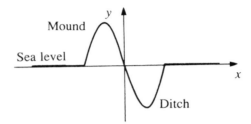

(a)   Find the area of the cross-section of the ditch which lies below sea-level.

(b)   Find the height of the top of the mound above sea-level, to the nearest centimetre.

[SMP 1991]

7   (a)   Use the trapezium rule with 5 ordinates to find

$$\int_0^1 \frac{2}{1+x^2} \, dx$$

(b)   State the approximate value of

$$\int_{-1}^1 \frac{2}{1+x^2} \, dx$$

8   The graph of the function $f(x) = x^2 + x - 2$ is shown.

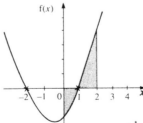

(a)   Find $\int f(x) \, dx$. Hence evaluate $\int_0^1 f(x) \, dx$ and $\int_1^2 f(x) \, dx$.

(b)   Find the area shaded.

[SMP 1990]

9   Use a numerical method to estimate the gradient of the graph of $y = 2^x$ at the point (3, 8).

FUNCTIONS

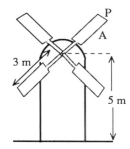

1 On the windmill, the point P is at the middle of the tip of the sail A. At time $t = 0$, A is horizontal and the sails make a complete turn every 2 minutes.

An expression, in terms of t, for the height (h metres) of P above the ground is:

$$h = a + b \sin \pi t$$

(a) Write down the values of a and b.

(b) Work out the first time at which P is 6 metres above the ground.

2 Solve, for $0 \le \theta \le 360°$:

(a) $4 \cos (\frac{1}{2}\theta + 10°) = 3$

(b) $9 \sin^2 \theta = 1$

3 The approximate depth of water, in metres, at the entrance to Eastminster harbour t hours after noon, is given by:

$$2.3 \cos (0.5t) + 6.1$$

(a) Calculate the time when the depth is first equal to 4 metres.

(b) Calculate the greatest rate at which the tide falls.

4 The Big Top of a circus is a circular tent of radius 20 metres.

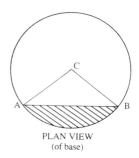

PLAN VIEW
(of base)

(a) A banner 30 metres long is stretched around a section AB of the perimeter of the base of the tent. Find, in radians, the angle ACB where C is the centre of the circular base shown in the plan view.

(b) One evening, the seats in the shaded segment are reserved for a large school party. Given that the density of seating is 1.2 children per square metre, find the number of children in the school party.

[SMP 1991]

5 £1000 is invested in a savings account. The account earns interest and, after n years, the amount of money in the account is £T, where

$$T = 1000 \times 1.1^n$$

(a) State the annual rate of interest.

(b) Show that $n = \dfrac{\log_{10} T - 3}{\log_{10} 1.1}$.

(c) Calculate how many years it will be before the amount of money has doubled.

6 A fitness fanatic is being timed as she rides on an exercise bicycle. She starts, at time $t = 0$, with one pedal at 60° in front of the upward vertical. She then pedals at a constant speed.

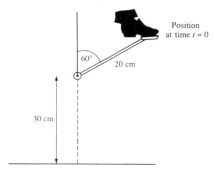

(a) Given that the height h of this pedal above the floor at time t is given by:

$$h = a + b \cos (180 (t + c)°)$$

where t is measured in seconds and a and b in centimetres, use the information in the diagram to find the values of a, b and c.

(b) Find the times, to the nearest 0.1 of a second, at which the height of the pedal above the floor is 45 centimetres, for $0 \le t \le 4$.

[SMP 1991]

7 Find the equation of the tangent to the curve $y = e^{kx}$ at the point $(\frac{1}{k}, e)$, where $k \ne 0$, giving your answer in the simplest form.

[SMP 1981]

8 A large waste skip is as illustrated. The area of cross section is a trapezium, where θ is the angle the two sloping sides make with the vertical.

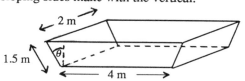

(a) Show that the volume V of the skip is such that

$$V = 12 \cos \theta + 4.5 \sin \theta \cos \theta$$

(b) Find the value of θ (to 2 d.p.) which maximises the volume of the skip. Hence find the maximum volume of the skip.

MATHEMATICAL METHODS

1   The position of a moving point is given by the equations

$$x = 3 + 5 \cos t, \quad y = 4 + 5 \sin t$$

(a)   Express $\cos t$ in terms of x and $\sin t$ in terms of y.

(b)   Use the identity $\sin^2 t + \cos^2 t = 1$ to find an equation of the path of the point in terms of x and y.

(c)   Explain why the path is a circle and give its radius and centre.

2   ABCD is a parallelogram whose diagonals intersect, at angle $\theta°$, at the point E.

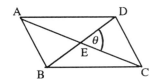

Let $AB = CD = x$, $AD = BC = y$, $BD = p$, $AC = q$

(a)   Apply the cosine rule to triangle DEC to show that
$$x^2 = \frac{1}{4}p^2 + \frac{1}{4}q^2 - \frac{1}{2}pq \cos \theta$$

(b)   By using triangle DEA, show that
$$y^2 = \frac{1}{4}p^2 + \frac{1}{4}q^2 - \frac{1}{2}pq \cos (180 - \theta)$$

(c)   By referring to a cosine graph, explain why
$$\cos (180 - \theta) = -\cos \theta$$

(d)   Using these three results, show that
$$2(x^2 + y^2) = p^2 + q^2$$

(e)   A parallelogram has sides of 3 cm and 5 cm and one diagonal of 6 cm. Calculate the length of the other diagonal.

3   A function called **soc** has the property that, if $y = \text{soc } x$, then $\frac{dy}{dx} = \sqrt{(1 + x^2)}$. Use the chain rule to find expressions for $\frac{dy}{dx}$ if

(a) $y = \text{soc } (2x)$ (b) $y = \text{soc } (\sqrt{x})$, where $x > 0$

[SMP 1983]

4   By using the expression $r \cos (\theta + \alpha)$, solve

$$5 \cos \theta - 2 \sin \theta = 0.6 \quad \text{for } 0° \le \theta \le 360°$$

By drawing the graphs of $y = 5 \cos \theta - 2 \sin \theta$ and $y = 0.6$, check that you have found all the solutions.

5   Prove that the two lines with equations

$$\mathbf{r} = \begin{bmatrix} 0 \\ 2 \\ -3 \end{bmatrix} + s \begin{bmatrix} 1 \\ -1 \\ -1 \end{bmatrix} \quad \text{and} \quad \mathbf{r} = \begin{bmatrix} -1 \\ 6 \\ -1 \end{bmatrix} + t \begin{bmatrix} 2 \\ 1 \\ -1 \end{bmatrix}$$

have a point in common.

Find, in parametric form, an equation for the plane containing the two lines.

[SMP 1978]

6   Find the binomial approximation, as far as the term in h^2, for $\sqrt{(1 + h)}$ when h is small. Use this to find an approximate value for

$$\int_0^{0.04} \sqrt{(1 + \sqrt{x})} \, dx$$

rounding your answer to 4 decimal places.

[SMP 1980]

7   A 'robot arm' signwriter is programmed to write a large letter V onto a piece of plywood. From its starting position at the origin, it moves through the vector $\begin{bmatrix} 30 \\ 50 \\ 80 \end{bmatrix}$ to the top left point of the V. The V is drawn by the vector $\begin{bmatrix} 5 \\ 3 \\ 1 \end{bmatrix}$ followed by $\begin{bmatrix} 3 \\ -5 \\ -1 \end{bmatrix}$.

(a)   What is the angle between the two arms of the V?

(b)   What is the vector equation of the plane of the plywood?

(c)   What vector will take the arm back to its starting position?

8   In a simple model of an epidemic, the rate of spread is proportional both to the number y who have been infected and to the number who remain uninfected.

(a)   If the total population is N, write down a differential equation for the number infected after time t.

(b)   In a population of 500, 1 person is infected initially and a week later 6 are infected. Find the differential equation that describes this particular epidemic.

(c)   Use a step-by-step method with step length 1 to estimate the number infected after 3 weeks.

9   (a)   For $x > 0$, sketch the graphs of $\frac{1}{x}$ and $\ln x$ on the same axes.

(b)   Sketch the graph of $S = \frac{1}{x} + \ln x$ against x.

(c)   Find the minimum value of S.

CALCULUS METHODS

1 Find (a) $\int \dfrac{2x-1}{x^3}\,\mathrm{d}x$ (b) $\int_1^{10} \dfrac{(1-s^2)^2}{s^2}\,\mathrm{d}s$

     (c) $\int_1^4 \dfrac{1}{\sqrt[3]{x}}\,\mathrm{d}x$

2 Find the positive value of x for which $\dfrac{x}{(3x+2)^2}$ takes
its greatest value. (You need **not** prove it is a maximum
rather than a minimum.)

[SMP 1982]

3 Find the equation of the tangent to the curve $y = x^2 \ln x$
at the point (e, e^2).

[SMP 1982]

4 A teapot is cylindrical in shape
with radius r and height h cm.
It is designed to hold 1500 cc.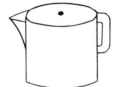

  (a) Find h in terms of r.

  (b) Hence find an expression for the total surface
area A (taken to be a cylinder) in terms of r, and
show that this simplifies to

$$A = \frac{3000}{r} + 2\pi r^2$$

  (c) The loss of heat, L, from a teapot is proportional to
its surface area, A,

    i.e. $L = kA$ where k is a constant

Write down an expression for L and use it to find
values of r and h which will minimise the loss of
heat from the pot.

  (d) Another teapot with capacity
1500 cc can be taken to be
spherical in shape. By comparing surface areas, decide
whether it will retain heat
better than the cylindrical
teapot.

    (Volume of sphere $= \dfrac{4}{3}\pi r^3$;

    surface area of sphere $= 4\pi r^2$)

5 The position vector of a particle after t seconds is:

$$\begin{bmatrix} 20t \\ 12t - 5t^2 \end{bmatrix}$$

  (a) Find the velocity, $\mathbf{v}$, in terms of t.

  (b) Find the value of t when the **speed** is least and
state this minimum speed.

6 (a) Explain why the volume of the cone can be
thought of as

$$\pi \int_0^h (y \tan 30°)^2 \,\mathrm{d}y$$

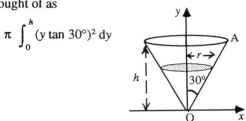

  (b) Find the volume in terms of h.

7 (a) For the curve with equation $(x - y)^2 = x + y$, find
$\dfrac{\mathrm{d}y}{\mathrm{d}x}$ in terms of x and y.

  (b) Find the gradient of the curve at the points where
it cuts the axes.

  (c) Sketch a part of the curve including these points.

8 Find the volume of the solid of revolution formed by
rotating the graph of $y = \dfrac{1}{(2x+1)}$, from $x = 0$ to $x = 2$,
about the x-axis.
[You may leave your answer as a multiple of π]

[SMP 1985]

9 Using the substitution $3x + 2 = u$, or otherwise,
evaluate

$$\int_0^8 \frac{x}{(3x+2)^2}\,\mathrm{d}x$$

giving your answer correct to 3 decimal places.

10 The following steps give a method for finding a numerical approximation to π. You are asked to carry them
out.

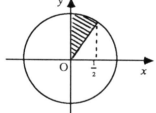

  (a) The diagram represents a circle of unit radius.
Show that the shaded area is $\dfrac{1}{12}\pi$ square units.

  (b) Deduce that: $\pi = 12 \displaystyle\int_0^{1/2} \sqrt{(1-x^2)}\,\mathrm{d}x - \dfrac{1}{2}\sqrt{27}$

  (c) Give the Maclaurin's series for $\sqrt{(1 + z)}$, neglecting powers of z above the second. Deduce an
approximation for $\sqrt{(1 - x^2)}$, neglecting powers of x
above the fourth.

  (d) Use your approximations to estimate both

$$\int_0^{1/2} \sqrt{(1 - x^2)}\,\mathrm{d}x \text{ and } \sqrt{(25 + 2)}$$

  (e) Hence estimate π, giving your answer to 3 s.f.

[SMP 1977]

SOLUTIONS

FOUNDATIONS

Graphs

1 $5y + 2x = 23$

2 (a)

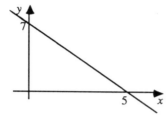

(b)

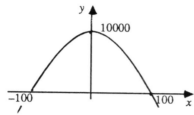

3 (a) $3y + 5x = 15$

* (b) $y = (x - 2)^2 + 3$

* (c) $y = -(x + 3)(x - 1)$

[* other answers are possible]

4 (a) (i) $(x + 3)^2 - 16$ (ii) $(x - 1)(x + 7)$

(b)

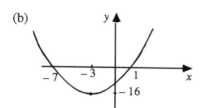

(c) $x = -3$

5 (a) $(x - 4)^2 + 9$

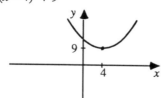

(b) $y = x^2 + 8x - 25$ has no zeros.

6 $a = 1, b = 5, c = 10$

---

7 (a) $(x + 1)(x + 7)$ (b) $(x - 7)(x - 3)$

(c) $(x + 1)(x + 5)$ (d) $(x - 7)^2$

(e) $(p - 4)(p + 4)$ (f) $(t - 5)(t + 3)$

8 (a) $0, 8$ (b) $-7, 1$ (c) -2

(d) $1, 9$ (e) $-1, -3$ (f) $-4, 3$

Sequences

1 (a) 1392 (b) -2415

(c) $2^{21} - 1 = 2\,097\,151$ (d) 25.645

2* (a) $\displaystyle\sum_{i=1}^{19} \frac{1}{3i - 1}$ (b) $\displaystyle\sum_{i=1}^{87} \frac{i}{i + 1}$

(c) $\displaystyle\sum_{i=1}^{50} (2i)^3$

[* other answers are possible]

3 (a) $-1, -3, -5, -7, -9$
$u_{30} = -59$

(b) $4, 8, 16, 32, 64$
$u_{30} = 2^{31}$

(c) $5, \dfrac{1}{5}, 5, \dfrac{1}{5}, 5$

$u_{30} = \dfrac{1}{5}$

4 (a) $u_2 = 2, u_4 = 24$

(b) With the $n + 1$th object in any particular position, there are u_n arrangements of the others.

There are $n + 1$ possible placings of the $n + 1$th object, giving $(n + 1)\, u_n$ arrangements in total.

(c) $u_{10} = 10 \times 9 \times 8 \times 7 \times 6 \times 5 \times 4 \times 3 \times 2 \times 1$
$= 3\,628\,800$

5 $\dfrac{1}{1 - \frac{1}{2}} = 2$

6 $5\left(\dfrac{5^{2n} - 1}{5 - 1}\right) = \dfrac{5}{4}(5^{2n} - 1)$

7 $225 + 234 + \ldots + 540$

$= 36 \times \dfrac{225 + 540}{2}$

$= 13\,770$

8 (a) $k = 5, n = 50$ (b) 4075

9 (a) $\dfrac{7}{1 - \frac{1}{2}} = 14$ (b) $\dfrac{3}{1 + \frac{2}{3}} = \dfrac{9}{5}$

10 (a) $4(n - 1)$, for $n \geq 2$

(b) The $n + 1$th diamond consists of its perimeter, with $4n$ dots, surrounding the nth diamond.

(c) $d_1 = 1, d_2 = 1 + 4, d_3 = 1 + 4 + 8$,
$d_4 = 1 + 4 + 8 + 12, \ldots$

So $d_n = 1 + 4\,(1 + 2 + \ldots + (n - 1))$
$ = 1 + 4 \times \dfrac{1}{2}\,(n - 1)n$
$ = 2n^2 - 2n + 1$

Functions and graphs

1 (a) (i) 80.5 (ii) 3 (iii) 1.789

(b) $x \in \mathbb{R}^+$

2 $x = 0, 3$

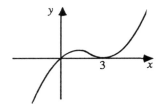

3

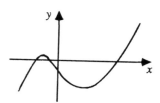

For large x, $f(x) \approx x^3$
For small x, $f(x) \approx -42x - 27$
The zeros are at $-1.5, -1$ and 9.

4 (a) $y = |x - 2| - 1$

(b) $y = \sqrt{(x + 1)} + 2$

5 (a) $(x - 2)^3 + 9$

(b) $g(x)$ is a translation of $f(x)$ through $\begin{bmatrix} 2 \\ 9 \end{bmatrix}$.

(c)

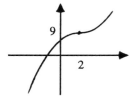

6 (a) $-ax^3, a > 0$ (b) $bx^2, b > 0$

(c) $-x^3 + 3x^2$ (d) $-2x^3 + 6x^2$

Expressions and equations

1 (a) $4xy$ (b) a^2

2 (a) $-2, 4$ (b) $\dfrac{1}{2}(7 \pm \sqrt{37})$

3 $-1 < x < 6$

4 (a)

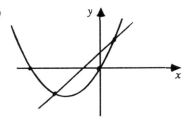

(b) $ x^2 + 7x = 4x + 4$
$\Rightarrow x^2 + 3x - 4 = 0$
$\Rightarrow x = -4 \text{ or } 1$

Therefore $-4 < x < 1$

5 (a) $72, 0, 0, 0$

(b) $3(x - 2)(x - 1)\,x(x + 1)$

6 $P(x) = (x - 5)(x^2 - 4) = (x - 5)(x - 2)(x + 2)$
$-2 \leq x \leq 2 \text{ or } x \geq 5$

7 (a) $x \pm 1, x \pm 5$

(b) $(x - 1)(x^2 + 7x + 5)$

(c) $1, -0.807, -6.19$

8 (a) $(x + 3)(x + 2)(x - 4)$ (b) $-3, -2, 4$

(c) $-$ (d) $-3 \leq x \leq -2 \text{ or } x \geq 4$

9 $(x + 3)^2 - 9 - 5 = 0$
$(x + 3)^2 - 14 = 0$
$(x + 3)^2 = 14$
$ x + 3 = \pm\sqrt{14}$
$ x = -3 \pm \sqrt{14}$

Numerical methods

1 (a) $-3 < x < -2,\ 2 < x < 3$

  (b) $-2 < x < -1,\ 0 < x < 1,\ 2 < x < 3$

  (c) $-2 < x < -1,\ 1 < x < 2$

2 (a) $6x = 8 - x^3$

  $\Rightarrow x = \frac{1}{6}(8 - x^3)$

  (b) $x^3 = 2x - 5$

  $\Rightarrow x = \frac{2x - 5}{x^2}$

  (c) $\sqrt{x} = 3 - 2x$

  $\Rightarrow x = (3 - 2x)^2$

  (d) $x^2 = \frac{1}{x} - 2$

  $\Rightarrow x = \frac{1}{x^2} - \frac{2}{x}$

3 (a) $-2.2,\ -1,\ 3.2$

  (b) $(x + 1)(x^2 - x - 7)$

  (c) $-1,\ \frac{1}{2}(1 \pm \sqrt{29})$

4

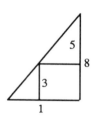

$1 + \frac{3}{8} = 1.375$

5 (a)

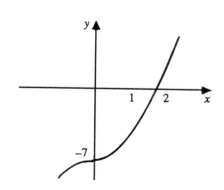

$1 < x < 2$

  (b) $x^3 = 7 \Rightarrow x^4 = 7x$

  $\Rightarrow x = \sqrt{\sqrt{(7x)}}$

  (c) 1.913

6 (a)

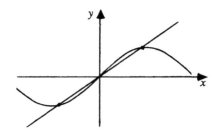

There are 3 roots.

  (b) $-1.496,\ 0,\ 1.496$

INTRODUCTORY CALCULUS

Rates of change/Gradient of curves

1 (a) $-\frac{1}{2}$ (b) $2s = 4 - t$

2 $p = 2r + 1$

3 (a) $u = 15 + 6t,\ v = 10 + 8t$

  (b) Firm B is the cheaper for any time up to $2\frac{1}{2}$ hours, the charges for both firms are the same for $2\frac{1}{2}$ hours and thereafter firm A is the cheaper.

  (c) $\frac{du}{dt} = 6$ and $\frac{dv}{dt} = 8$

  $\frac{du}{dt}$ is the extra charge per hour by firm A.

  $\frac{dv}{dt}$ is the extra charge per hour by firm B.

  (d) $C = 4u + 3v$

  $\frac{dC}{dt} = 4 \times 6 + 3 \times 8 = 48$, which is the cost per hour for the group of 4 plumbers from firm A and 3 from firm B.

4 (a)

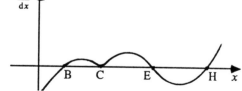

  (b) (i) B, C, E, H   (ii) B, E, H

    (iii) B, H   (iv) J   (v) G, I

5 (a) $15x^2 + 1$ (b) $-1 + 6x$

6 $y = 9x - 11$

Optimisation

1

	A	B	C	D	E
f(x)	–	+	+	+	–
f′(x)	+	+	–	–	–

2 B(max), D(min)

3 $v = 3$ when $t = 1$

4 (a) $(0, -1)$ is a min.

(b) $(0, 0)$ is a min. , $(-2, 0.54)$ is a max.

5 min. $(-3, -123)$, max $(0, 12)$, min. $(1, 5)$

6 $V = \pi r^2 h = \pi r^2(6 - 2\pi r)$
$$= 6\pi r^2 - 2\pi^2 r^3$$

$$\frac{dV}{dr} = 0 \Rightarrow 12\pi r - 6\pi^2 r^2 = 0$$
$$\Rightarrow r = 0 \text{ or } \frac{2}{\pi}$$

$$V \max = \pi \left(\frac{2}{\pi}\right)^2 \left[6 - 2\pi\left(\frac{2}{\pi}\right)\right]$$

$$\approx 2.55 \text{ m}^3$$

7 (a) The base of the box has area $(12 - 2x)^2$ cm².

The box has height x cm.

$$V = x (12 - 2x)^2 = 4x (6 - x)^2$$

(b) $V = 144x - 48x^2 + 4x^3$

$$\frac{dV}{dx} = 0 \Rightarrow 144 - 96x + 12x^2 = 0$$
$$\Rightarrow x = 2 \text{ or } x = 6$$

V max occurs when $x = 2$ and the box is then 8 cm x 8 cm x 2 cm.

Numerical integration

1 (a) $C = 2\pi r \Rightarrow r = \dfrac{C}{2\pi}$

$$A = \pi r^2 = \pi\left(\frac{C}{2\pi}\right)^2 = \frac{C^2}{4\pi}$$

h	0	1	2	3	4	5	6
A	0.509	0.389	0.275	0.154	0.086	0.036	0.014

(b) (i) 1.158 m³   (ii) 1.202 m³

2 (a) 8.251   (b) 8.251

3 (a) 2.492 (using the trapezium rule with 4 strips)

(b) This is an over-estimate since the curve is concave.

4 0

5 $\dfrac{2\pi}{3}$

6 (a) B + C   (b) A + B   (c) A + B + C

7 Cross-sectional area ≈ 0.06 m² (using the trapezium rule with 8 strips)

Mass $\approx 8 \times 4 \times 0.06 = 1.92$ tonnes

Algebraic integration

1 (a) $-6\dfrac{3}{4}$

(b) The curve lies below the x-axis.

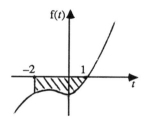

2 4.117

3

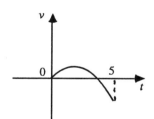

$$\int_2^3 (12t - 3t^2) \, dt = \left[6t^2 - t^3 \right]_2^3 = 11 \text{ m}$$

4 (a) $\dfrac{x^4}{4} - x^2 + c$

(b) $3x^3 + 6x^2 + 4x + c$

5 $y = x^2 - \dfrac{x^3}{6}$

6 (a) $\dfrac{5}{6}$ (b) 33.6

7 $2t + 2$

8 (a) $21\dfrac{1}{3}$ (b) $11\dfrac{1}{4}$

FUNCTIONS

Algebra of functions

1 (a) (i) $gf(-1) = g(0) = 1$ (ii) $gf(x) = \cos(x^3 + 1)$

  (b) $fg(x) = \dfrac{3 - \frac{1}{x}}{\frac{1}{x} + 1} = \dfrac{3x - 1}{1 + x}$

  Domain is all real numbers except -1.
  Range is all real numbers except 3.

2 Use flow diagrams for (a) and (b):

  (a) $f^{-1}(x) = \sqrt{\left(\dfrac{x-4}{3}\right)}$ (b) $f^{-1}(x) = \sqrt[3]{x} - 2$

  (c) Put $y = f(x)$:

$$\Rightarrow y(5x - 3) = 2x + 1$$
$$\Rightarrow 5xy - 2x = 3y + 1$$
$$\Rightarrow x = \dfrac{3y + 1}{5y - 2}$$
$$\Rightarrow f^{-1}(x) = \dfrac{3x + 1}{5x - 2}$$

3 (a) $y(x + b) = a$

   $yx + yb = a$

   $x = \dfrac{(a - yb)}{y}$

  (b) $t(a - b) = a$
   $at - tb = a$
   $at = a + tb$

   (i) $tb = at - a$
    $b = \dfrac{at - a}{t}$

   (ii) $a(t - 1) = tb$
    $a = \dfrac{tb}{t - 1}$

4 (a) $N = 20(C - 9)$

  (b) $C = \dfrac{Nx}{100} + y$

5 (a)

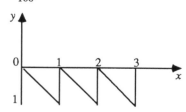

  (b)

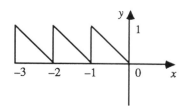

(c)

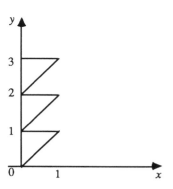

(d)

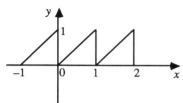

Circular functions

1 (a) $y = 3 \sin 4x$ (b) $x = 2 \sin(t + 30)°$

2 $-306°, -54°, 306°$

3 A $(0°, 2.6)$, B $(15°, 3)$, C $(60°, 0)$, D $(105°, -3)$,
  E $(150°, 0)$

4 $\sin x = 0.667 \Rightarrow x = 41.8°, 138.2°, 401.8°, 498.2°,$
  $761.8°, 858.2°$

  $4t + 30 = x \Rightarrow t = 3, 27, 93, 117$

5 $\dfrac{\sin x}{\cos x} = \dfrac{3}{2} \Rightarrow \tan x = 1.5$
    $x = 56.3°, -123.7°$

6 (a) $y = 7.5$, $r = 6$

  (b) $6 = 7.5 - 6 \cos 8t°$
   $\cos 8t° = 0.25$

  (c) $8t = \cos^{-1} 0.25 \Rightarrow 8t = 75.5, 284.5, \ldots$

   The first two times are $\dfrac{75.5}{8} = 9.4$ seconds

   and $\dfrac{284.5}{8} = 35.6$ seconds.

Growth functions

1 (a) -2 (b) -1 (c) 1.5 (d) 1.5

2 (a) $2^6 = 64$, so $5t + 1 = 6$ and $t = 1$

  (b) $t - 2 = \dfrac{\log 12}{\log 5} \approx 1.54 \Rightarrow t = 3.54$ (to 3 s.f.)

3

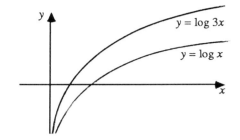

$\log 3x = \log 3 + \log x$ and so the transformation is a translation with vector $\begin{bmatrix} 0 \\ \log 3 \end{bmatrix}$.

4 Rate of growth $= \dfrac{17\,500}{14\,000} = 1.25$

Population after three years $= 14\,000 \times 1.25^3 \approx 27\,000$

5 (a) $0.72 = ka^0$, so $k = 0.72$

$0.37 = ka^{26}$

$\Rightarrow a^{26} = \dfrac{0.37}{0.72}$

$\Rightarrow a = \left(\dfrac{0.37}{0.72}\right)^{\frac{1}{26}} = 0.975$ (to 3 s.f.)

(b) When 75% have use of a car, $P = 0.25$

$0.25 = 0.72 \times 0.975^t$

$0.975^t = \dfrac{0.25}{0.72} = 0.347$

$t = \dfrac{\log 0.347}{\log 0.975} = 41.8$

At least 75% will have use of a car 42 years after 1961, in 2003.

6 (a) $400 \times 1.7^{12} = 233000$ (to 3 s.f.)

(b) $1.7^t = 2 \Rightarrow t = 1.306$
1 hour 18 minutes

(c) $1\,000\,000T = 400 \times 1.7^T$

$T = \dfrac{\log 2500T}{\log 1.7}$

(d) $T = 20.43$ (to 2 d.p.)

7 (a) $\dfrac{370}{100} = 3.7$

(b) (i) $3.7^{\frac{1}{12}} \approx 1.1152$

(ii) $£100 \times 1.1152^5 \approx £172.50$

Radians

1 (a) $\dfrac{180}{3} = 60°$ (b) $7 \times \dfrac{180}{8} = 157.5°$

(c) $0.3 \times \dfrac{180}{\pi} = 17.2°$ (to 1 d.p.)

2 (a) $\dfrac{\pi}{4}$ (b) $100 \times \dfrac{\pi}{180} = 1.745$

(c) $\dfrac{5\pi}{4}$ (d) $280 \times \dfrac{\pi}{180} = 4.887$

3 (a) Area of bed $= \dfrac{1}{2} \times 16 \times 1.4 = 11.2$ m²

$= 11.2 \times 100^2$ cm²

Number of plants $= 11.2 \times 100^2 \div 300 = 373$

(b) Length of curved edge $= 4 \times 1.4 = 5.6$ m
Length of edging strip $= 5.6 + 8 = 13.6$ m

4 (a) Amplitude 10 m, period 10 seconds

(b) $10 \sin \dfrac{\pi t}{5} = -5$

$\Rightarrow \sin \dfrac{\pi t}{5} = -0.5$

$\Rightarrow \dfrac{\pi t}{5} = \dfrac{7\pi}{6}, \dfrac{11\pi}{6}$

$t = \dfrac{35}{6}, \dfrac{55}{6}$

The crests of the waves are above the deck for $\dfrac{20}{6} = 3\dfrac{1}{3}$ seconds of each cycle.

(c) $\dfrac{dh}{dt} = 2\pi \cos \dfrac{\pi t}{5}$

When the rate of descent is 4 m s⁻¹, $\dfrac{dh}{dt} = -4$

$2\pi \cos \dfrac{\pi t}{5} = -4$

$\cos \dfrac{\pi t}{5} = -\dfrac{4}{2\pi}$

$\dfrac{\pi t}{5} = 2.26, 4.02$

$t = 3.6$ and 6.4 seconds

Between these times the rate of descent is more than 4 m s⁻¹, so this occurs for 2.8 seconds per cycle.

Percentage of cycle $= \dfrac{2.8}{10} \times 100 = 28\%$

5 The principal value of $\tan^{-1} \pi = 1.26$
Other values are $1.26 \pm \pi$, $1.26 \pm 2\pi$, etc.

6 (a) 10θ (b) $5 \sin \theta$

(c) $p = $ arc AB $+ 2 \times$ AC $= 10\theta + 10 \sin \theta$
$= 10(\theta + \sin \theta)$

(d) $\dfrac{dp}{d\theta} = 10 + 10 \cos \theta$

When $\theta = 0$, $\dfrac{dp}{d\theta} = 20$

7 (a) $42 - 2r = r\theta \Rightarrow \theta = \dfrac{42}{r} - 2$

$A = \dfrac{1}{2}r^2\theta = 21r - r^2$

(b) $\dfrac{dA}{dr} = 21 - 2r$ so $\dfrac{dA}{dr} = 0$ when $r = 10.5$

Then $A = 110.25$ cm².

Solutions

e

1 (i) C (ii) A (iii) B

2 (a) 3 (b) $\frac{1}{2}$ (c) −1

3 $\left[\frac{1}{3}e^{3x}\right]_1^2 \approx 127.8$

4 (a) $N = \dfrac{500}{1+249\,e^{-1}} = 5.4$

5 students have heard the rumour after 1 hour.

(b) $250 = \dfrac{500}{1+249\,e^{-t}}$

$249\,e^{-t} = 1$

$-t = \ln\dfrac{1}{249} = -5.5$

Half the students have heard the rumour after $5\frac{1}{2}$ hours.

(c) $499 = \dfrac{500}{1+249\,e^{-t}}$

$249\,e^{-t} = \dfrac{500}{499} - 1 = 0.002\,004$

$-t = \ln\dfrac{0.002\,004}{249} = -11.73$

All the students except one have heard the rumour after approximately $11\frac{3}{4}$ hours.

5 $\dfrac{dy}{dx} = \dfrac{1}{x}$, so the gradient of the tangent at $x = 0.5$ is 2.

At $x = 0.5$, $y = \ln 0.5 = -0.69$ (to 2 s.f.)

$\dfrac{y+0.69}{x-0.5} = 2 \Rightarrow y = 2x - 1.69$

6 (a) $m = 3e^{-4} = 0.055$ kg (to 2 s.f.)

(b) $1.5 = 3e^{-0.4t}$

$\Rightarrow -0.4t = \ln 0.5$

$\Rightarrow t = \dfrac{\ln 0.5}{-0.4} = 1.73$ hours

The mass is half its initial value after $1\frac{3}{4}$ hours.

(c) $\dfrac{dm}{dt} = -1.2\,e^{-0.4t}$

At $t = 4$, $\dfrac{dm}{dt} = -1.2e^{-1.6} = -0.24$

The mass is decaying at 0.24 kg per hour.

Transformations

1 The translation has vector $\begin{bmatrix} 0 \\ \ln 0.5 \end{bmatrix}$.

The stretch has scale factor 2 from the *y*-axis.

2 (a) $y = 4\cos(2x - 0.8)$

(b) $y = 4\sin(2x + \frac{\pi}{2} - 0.8)$ or $y = 4\sin(2x + 0.77)$

3 (a) A translation of $\begin{bmatrix} 5 \\ -9 \end{bmatrix}$ gives $y + 9 = \dfrac{1}{x-5}$

Reflection in the *x*-axis replaces *y* by − *y*,

$-y + 9 = \dfrac{1}{x-5}$

$y = 9 - \dfrac{1}{x-5} = \dfrac{9(x-5)-1}{x-5} = \dfrac{9x-46}{x-5}$

(b) $y = \dfrac{1}{x} \rightarrow y = -\dfrac{1}{x} \rightarrow y + 3 = -\dfrac{1}{x-2}$

$y = \dfrac{3x-5}{-x+2}$

4

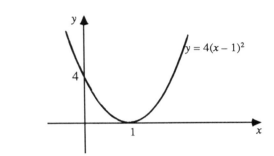

5 It looks as if the graph of $y = \dfrac{-1}{x}$ has been translated by $\begin{bmatrix} 4 \\ -1 \end{bmatrix}$.

This gives the equation:

$y + 1 = \dfrac{-1}{x-4}$

$y = \dfrac{-1}{x-4} - 1 = \dfrac{-1-1(x-4)}{x-4} = \dfrac{3-x}{x-4}$

6 (a) The equation is $(\frac{1}{3}x)^2 + (\frac{1}{2}y)^2 = 1$

or $\dfrac{x^2}{9} + \dfrac{y^2}{4} = 1$

(b) $\left(\dfrac{x-2}{4}\right)^2 + \left(\dfrac{y-1}{3}\right)^2 = 1$

MATHEMATICAL METHODS

The power of Pythagoras

1 $\sin 30° = \dfrac{1}{2}$, $\cos 30° = \dfrac{\sqrt{3}}{2}$, $\tan 30° = \dfrac{1}{\sqrt{3}}$

$\sin 60° = \dfrac{\sqrt{3}}{2}$, $\cos 60° = \dfrac{1}{2}$, $\tan 60° = \sqrt{3}$

$\sin 45° = \dfrac{1}{\sqrt{2}}$, $\cos 45° = \dfrac{1}{\sqrt{2}}$, $\tan 45° = 1$

66

2 (a) $(x - 4)^2 + (y - 3)^2 = 25$

Intersects x-axis when $y = 0$.
Then $(x - 4)^2 = 16 \Rightarrow x - 4 = \pm 4$
intersects x-axis at $(0, 0)$ and $(8, 0)$.
Similarly, intersects y-axis at $(0, 0)$ and $(0, 6)$.

(b) $(x - 1)^2 + (y + 2)^2 + (z + 3)^2 = 25$

3 (a) $2 \cos^2 x + 3 \sin x = 0$
$2 (1 - \sin^2 x) + 3 \sin x = 0$
$(2 \sin x + 1)(\sin x - 2) = 0$
$\sin x = -\dfrac{1}{2} \Rightarrow x = \dfrac{7\pi}{6}, \dfrac{11\pi}{6}$

(b) $\sin 2x = \cos x$
$2 \sin x \cos x - \cos x = 0$
$\cos x (2 \sin x - 1) = 0$
$\cos x = 0 \Rightarrow x = \dfrac{\pi}{2}, \dfrac{3\pi}{2}$
or $\sin x = \dfrac{1}{2} \Rightarrow x = \dfrac{\pi}{6}, \dfrac{5\pi}{6}$
$x = \dfrac{\pi}{6}, \dfrac{\pi}{2}, \dfrac{5\pi}{6}, \dfrac{3\pi}{2}$

4 (a) $r = \sqrt{(3^2 + 5^2)} = \sqrt{34} = 5.83$
$\alpha = \tan^{-1} \dfrac{5}{3} = 59.0°$
So $3 \sin \theta + 5 \cos \theta = 5.83 \sin (\theta + 59.0°)$
A sine wave, amplitude 5.83, translated 59° to the left.

(b) $5.83 \sin (\theta + 59.0°) = 4$
$\sin (\theta + 59.0°) = 0.686$
$\theta + 59.0° = 43.3°, 136.7°, 403.3°, \ldots$
$\theta = 77.7°, 344.3°$

5 (a) $\cos (2A + A) = \cos 2A \cos A - \sin 2A \sin A$

(b) $\cos 3A = (2 \cos^2 A - 1) \cos A$
$\qquad\qquad\qquad - (2 \sin A \cos A) \sin A$
$= 2 \cos^3 A - \cos A - 2 \sin^2 A \cos A$
$= 2 \cos^3 A - \cos A - 2(1 - \cos^2 A) \cos A$
$= 4 \cos^3 A - 3 \cos A$

6 (a) $\dfrac{5}{\sin C} = \dfrac{4}{\sin 40°} \Rightarrow \sin C = \dfrac{\sin 40°}{4} \times 5$
$\qquad\qquad\qquad\qquad C = 53.5°$ or $126.5°$

(b)
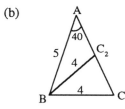
There are two possible triangles, as shown.

(c) $\sin C = \dfrac{\sin 40°}{6} \times 5 \Rightarrow C = 32.4°$ (or $147.6°$)

But $40 + 147.6 = 187.6$, which is impossible.
There is only one solution in this case.

7

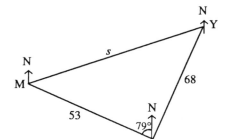

(a) $s = \sqrt{(68^2 + 53^2 - 2 \times 68 \times 53 \times \cos 100)}$
$= 93.2$ km

(b) $\dfrac{53}{\sin Y} = \dfrac{93.2}{\sin 100} \Rightarrow \sin Y = \dfrac{\sin 100}{93.2} \times 53$
$\qquad\qquad\qquad\qquad\qquad Y = 34.1°$

Bearing $= 180 + 21 + 34.1 = 235.1°$

Vector geometry

1 (a) $\mathbf{r} = \begin{bmatrix} -1 \\ 4 \end{bmatrix} + \lambda \begin{bmatrix} 5 \\ -6 \end{bmatrix}$

(b) $\begin{bmatrix} 1 \\ 5 \\ 0 \end{bmatrix} - \begin{bmatrix} -1 \\ 4 \\ 2 \end{bmatrix} = \begin{bmatrix} 2 \\ 1 \\ -2 \end{bmatrix} \Rightarrow \mathbf{r} = \begin{bmatrix} -1 \\ 4 \\ 2 \end{bmatrix} + \lambda \begin{bmatrix} 2 \\ 1 \\ -2 \end{bmatrix}$

2 (a) $\mathbf{r} = \begin{bmatrix} 0 \\ 1 \\ 1 \end{bmatrix} + s \begin{bmatrix} 2 \\ -1 \\ -1 \end{bmatrix}$ and $\mathbf{r} = t \begin{bmatrix} 3 \\ 0 \\ 1 \end{bmatrix}$

(b) At a point of intersection, $0 + 2s = 3t$ ①
$\qquad\qquad\qquad\qquad\qquad 1 - s = 0t$ ②
$\qquad\qquad\qquad\qquad\qquad 1 - s = t$ ③

① and ② give $s = 1$ and $t = \dfrac{2}{3}$, but these values of s and t do not satisfy ③, hence the lines do not intersect.

3 (a) (i) $\mathbf{a}.\mathbf{b} = 0 \Rightarrow \mathbf{a} = 0$ or $\mathbf{b} = 0$ or $\mathbf{a}$ is perpendicular to $\mathbf{b}$.

(ii) $\mathbf{a}.\mathbf{b} = ab \Rightarrow \mathbf{a}$ and $\mathbf{b}$ are parallel vectors.

(b) (i) $\mathbf{c}, \mathbf{d}; \mathbf{d}, \mathbf{e}$ \qquad\qquad (ii) $\mathbf{c}, \mathbf{e}$

4 (a) $\mathbf{r} = \begin{bmatrix} 5 \\ -1 \\ 6 \end{bmatrix} + \lambda \begin{bmatrix} 1 \\ 0 \\ 0 \end{bmatrix} + \mu \begin{bmatrix} -2 \\ 4 \\ 1 \end{bmatrix}$

(b) $x = 5 + \lambda - 2\mu$
$y = -1 + 4\mu$
$z = 6 + \mu$
Eliminating μ, $y - 4z = -25$.
(The equation is independent of x.)

5 (a) Let B be the angle between the normal vector of the plane and the x-axis.

$$\begin{bmatrix} 2 \\ -1 \\ 0 \end{bmatrix} \cdot \begin{bmatrix} 1 \\ 0 \\ 0 \end{bmatrix} = \sqrt{5}\,\sqrt{1}\cos B$$

Then $B = 26.6°$ and the angle between the plane and the x-axis is $90 - 26.6 = 63.4°$.

(b) The angle between the planes is equal to the angle between their normal vectors.

$$\begin{bmatrix} 2 \\ 1 \\ 0 \end{bmatrix} \cdot \begin{bmatrix} 3 \\ 0 \\ 2 \end{bmatrix} = \sqrt{5}\,\sqrt{13}\cos A \Rightarrow A = 41.9°$$

6 Substituting for x, y and z into the equation of the plane, $(2 + \lambda) - 3(1 - 4\lambda) + 3(\lambda) = 15 \Rightarrow \lambda = 1$. Substituting $\lambda = 1$ into the equation of the line gives a point of intersection at $(3, -3, 3)$.

7 (a) $$\begin{bmatrix} 0 \\ 8 \\ 0 \end{bmatrix} \cdot \begin{bmatrix} 4 \\ 4 \\ 6 \end{bmatrix} = \sqrt{64}\,\sqrt{68}\cos\theta \Rightarrow \theta = 60.9°$$

(b) $$\begin{bmatrix} 4 \\ 4 \\ 6 \end{bmatrix} \cdot \begin{bmatrix} -4 \\ -4 \\ 6 \end{bmatrix} = \sqrt{68}\,\sqrt{68}\cos\beta \Rightarrow \beta = 86.6°$$

(c) $\begin{bmatrix} 0 \\ 0 \\ 1 \end{bmatrix}$ is normal to ABCD, $\begin{bmatrix} 3 \\ 0 \\ -2 \end{bmatrix}$ is normal to ABE.

$$\begin{bmatrix} 0 \\ 0 \\ 1 \end{bmatrix} \cdot \begin{bmatrix} 3 \\ 0 \\ -2 \end{bmatrix} = \sqrt{1}\,\sqrt{13}\cos\phi \Rightarrow \phi = 123.7°$$

The angle between the two planes is $56.3°$.

(d) $\begin{bmatrix} 0 \\ 3 \\ 2 \end{bmatrix}$ is normal to BCE.

$$\begin{bmatrix} 3 \\ 0 \\ -2 \end{bmatrix} \cdot \begin{bmatrix} 0 \\ 3 \\ 2 \end{bmatrix} = \sqrt{13}\,\sqrt{13}\cos\alpha \Rightarrow \alpha = 107.9°$$

The angle between the two planes is $72.1°$.

8 (a) $\mathbf{r} = \begin{bmatrix} 3 \\ 5 \\ 2 \end{bmatrix} + \lambda \begin{bmatrix} 22 \\ 13 \\ 7 \end{bmatrix}$

(b) A normal vector to the plane is $\mathbf{n} = \begin{bmatrix} 22 \\ 13 \\ 7 \end{bmatrix}$, the direction of the laser beam. $\mathbf{r.n} = \mathbf{a.n}$ where $\mathbf{a}$ is the position vector of a point on the plane.

So $\begin{bmatrix} x \\ y \\ z \end{bmatrix} \cdot \begin{bmatrix} 22 \\ 13 \\ 7 \end{bmatrix} = \begin{bmatrix} 15 \\ 15 \\ 5 \end{bmatrix} \cdot \begin{bmatrix} 22 \\ 13 \\ 7 \end{bmatrix}$

$\Rightarrow 22x + 13y + 7z = 560$

(c) $22(3 + 22\lambda) + 13(5 + 13\lambda) + 7(2 + 7\lambda) = 560$
$\Rightarrow \lambda = 0.59$

The laser beam hits at $(16.0, 12.7, 6.1)$.

Binomials

1 (a) $a^6 + 6a^5b + 15a^4b^2 + 20a^3b^3 + 15a^2b^4 + 6ab^5 + b^6$

(b) $32p^5 - 480p^4q + 2880p^3q^2 - 8640p^2q^3 + 12\,960pq^4 - 7776q^5$

2 (a) 120 　　(b) $\dfrac{1000 \times 999 \times 998}{3 \times 2 \times 1} = 166\,167\,000$

3 $a = 9$, $b = 6$, $c = 10$

In Pascal's triangle, $\binom{10}{6}$ is found by adding the two numbers above it, $\binom{9}{5}$ and $\binom{9}{6}$.

From symmetry, $\binom{10}{6} = \binom{10}{4}$.

4 (a) $a^{20} + 20a^{19}b + 190a^{18}b^2 + 1140a^{17}b^3$

(b) $1024x^{10} - 5120x^7 + 11\,520x^4 - 15\,360x$

5 (a) $1 + (-\frac{1}{2})x + \dfrac{(-\frac{1}{2})(-\frac{3}{2})}{2!}x^2 + \dfrac{(-\frac{1}{2})(-\frac{3}{2})(-\frac{5}{2})}{3!}x^3$

$= 1 - \dfrac{x}{2} + \dfrac{3x^2}{8} - \dfrac{5x^3}{16}$ for $-1 < x < 1$

(b) Replacing x by $(-2x)$ in part (a):

$(1 - 2x)^{-1/2} \approx 1 + x + \dfrac{3x^2}{2} + \dfrac{5x^3}{2}$

for $-\dfrac{1}{2} < x < \dfrac{1}{2}$

(c) Substituting for $x = 0.01$ in part (b):

$\dfrac{1}{\sqrt{(0.98)}} \approx 1 + 0.01 + 0.00015 + \ldots = 1.010$ (to 3 d.p.)

6 (a) $(1 + x)^{-2} \approx 1 - 2x + 3x^2 - 4x^3 + 5x^4$ for $-1 < x < 1$

(b) $(1 - 3x)^{1/2} \approx 1 + (\frac{1}{2})(-3x) + \dfrac{(\frac{1}{2})(-\frac{1}{2})}{2}(-3x)^2$

$+ \dfrac{(\frac{1}{2})(-\frac{1}{2})(-\frac{3}{2})}{6}(-3x)^3 + \dfrac{(\frac{1}{2})(-\frac{1}{2})(-\frac{3}{2})(-\frac{5}{2})}{24}(-3x)^4$

$\approx 1 - \dfrac{3x}{2} - \dfrac{9x^2}{8} - \dfrac{27x^3}{16} - \dfrac{405x^4}{128}$ for $-\dfrac{1}{3} < x < \dfrac{1}{3}$

(c) $\dfrac{1}{5}(1 + 4x^2)^{-1/2} \approx \dfrac{1}{5}(1 + (-\frac{1}{2})(4x^2)$

$+ \dfrac{(-\frac{1}{2})(-\frac{3}{2})}{2}(4x^2)^2)$

$\approx \dfrac{1}{5}(1 - 2x^2 + 6x^4)$ for $-\dfrac{1}{2} < x < \dfrac{1}{2}$

7 Area of hole $= \pi\,(6 \pm 0.2)^2$
$= 36\pi\,(1 \pm 0.033)^2$
$\approx 36\pi\,(1 \pm 0.67)$
$\approx 113 \pm 7.5$ cm²

Resulting area $\approx (630 \pm 10) - (113 \pm 7.5)$
$\approx 517 \pm 17.5$ cm²

8 Volume of cube $= (5 \pm 0.1)^3$
$= 125(1 \pm 0.02)^3$
$\approx 125\,(1 \pm 0.06)$ cm³

Mass $= 1200 \pm 50 = 1200\,(1 \pm 0.042)$ g

Density $\approx \dfrac{1200\,(1 \pm 0.042)}{125\,(1 \pm 0.06)}$

$\approx 9.6\,(1 \pm 0.102)$
$\approx 9.6 \pm 1.0$ g cm⁻³

Chain rule

1 (a) $2 \sin x \cos x$ (b) $2x \cos x^2$

2 (a) $84x^2(4x^3 + 3)^6$ (b) $-3 \cos^2 x \sin x$

(c) $\dfrac{-8x^3}{(x^4 - 3)^3}$ (d) $\dfrac{1}{x}$

(e) $\dfrac{4}{4x - 1}$ (f) $2 \cos 2x\, e^{\sin 2x}$

(g) $\dfrac{x}{\sqrt{(x^2 - 1)}}$ (h) $30 \sin 3x \cos 3x$

3 (a) $\dfrac{1}{6}(2x + 5)^3 + c$ (b) $\left[\dfrac{1}{3}e^{3x}\right]_0^1 = 6.36$

(c) $-\dfrac{1}{5}\cos 5x + c$ (d) $\left[2 \sin \dfrac{1}{2}x\right]_0^\pi = 2$

(e) $3 \ln x + c$ (f) $\left[\dfrac{1}{4}(x^2 + 1)^2\right]_1^2 = 5\,\dfrac{1}{4}$

4 $h = 10r$ and $\dfrac{dh}{dt} = 0.7$ when $r = 2$

$V = \pi r^2 h = \dfrac{\pi h^3}{100} \implies \dfrac{dV}{dh} = \dfrac{3\pi h^2}{100}$

When $r = 2$, $h = 20$ and

$\dfrac{dV}{dt} = \dfrac{dV}{dh} \times \dfrac{dh}{dt} = 0.03\pi h^2 \times 0.7$

$= 26.4$ cm³ per month

5 $\displaystyle\int_0^{\pi/4} 2 \sin 4t\, dt = \left[-\dfrac{1}{2}\cos 4t\right]_0^{\pi/4} = 1$

6 (a) $\dfrac{dr}{dt} = 0.1t$ and $V = \dfrac{4}{3}\pi r^3 \implies \dfrac{dV}{dr} = 4\pi r^2$

(b) $\dfrac{dV}{dt} = \dfrac{dV}{dr} \times \dfrac{dr}{dt} = 4\pi r^2 \times 0.1t$

$= 22.6$ m³ per min after 3 min

7 (a) $f'(x) = \dfrac{2x}{x^2 + a}$

(b) $f'(x) = 0$ when $x = 0$,
giving stationary values
at $(0, \ln a)$.

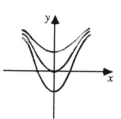

(c) $f(-x) = f(x)$ and so there is
symmetry about the y-axis.

8 (a) The cross-section is a trapezium.
$W = \dfrac{1}{2}(2 \sin x)(2 + 2 + 2 \cos x + 2 \cos x)$
$= \sin x\,(4 + 4 \cos x)$
$= 4 \sin x + 2 \sin 2x$

(b) $\dfrac{dW}{dx} = 4 \cos x + 4 \cos 2x$

At a stationary point:
$4 \cos x + 4\,(2 \cos^2 x - 1) = 0$
$2 \cos^2 x + \cos x - 1 = 0$
$(2 \cos x - 1)\,(\cos x + 1) = 0$
$\cos x = \dfrac{1}{2} \implies x = \dfrac{\pi}{3}$

Sketching the graph of W indicates that $x = \dfrac{\pi}{3}$
gives the maximum value for W.

Differential equations

1 (a) $y = x^3 + c$ (b) $y = \dfrac{1}{2}e^{2x} + c$

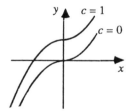
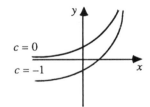

(c) $y = -\dfrac{1}{3}\cos 3x + c$

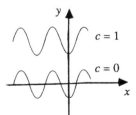

2 (a) $\dfrac{dy}{dx} = 2x + \dfrac{1}{x} \implies y = x^2 + \ln x + c$
$2 = 1 + \ln 1 + c \implies c = 1$
$y = x^2 + \ln x + 1$

(b) $y = \dfrac{1}{2}\sin(2x - 2) + c$

$y = \dfrac{1}{2}\sin(2x - 2) + 2$

3

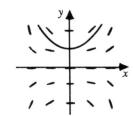

4

x	y	$\frac{dy}{dx}$	dx	dy	$x + dx$	$y + dy$
1	3.5	−0.5	0.2	−0.1	1.2	3.4
1.2	3.4	0.2	0.2	0.04	1.4	3.44
1.4	3.44				.	.
.	.				.	.
.	.				2.0	4.146 88

When $x = 2$, $y \approx 4.1$

5 $\quad \dfrac{dV}{dt} = -kV \Rightarrow V = Ae^{-kt}$ where V is the volume of water in the tank in litres and t is the time in minutes.

When $t = 0$, $V = 250$ and when $t = 4$, $V = 210$.

So $210 = 250e^{-4k} \Rightarrow 4k = \ln \left(\dfrac{25}{21}\right)$

$$\Rightarrow k = 0.04$$

$V = 250e^{-0.04t}$

6

x	y	$\frac{dy}{dx}$	dx	dy	$x + dx$	$y + dy$
1	2	3	0.2	0.6	1.2	2.6
1.2	2.6	3.6	0.2	0.72	1.4	3.32
1.4	3.32				.	.
.	.				.	.
.	.				2.0	6.46

When $x = 2$, $y \approx 6.5$

CALCULUS METHODS

Parameters

1 (a) The curve goes through the points:
(3, 0), (2, −2), (−1, −4), (−6, −6), (−13, −8)

(b) $y = -2t \Rightarrow t = -\dfrac{1}{2}y$

$x = 3 - (-\dfrac{1}{2}y)^2 = 3 - \dfrac{1}{4}y^2$ or $y^2 = 4(3 - x)$

The curve plotted in (a) is part of the graph of $y = -\sqrt{(4\,(3-x))}$.

2 The major axis is of length 10 and the minor axis of length 6.

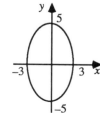

3 (a) $\dfrac{dy}{dx} = \dfrac{2 \cos \theta}{-3 \sin \theta}$

(b) At the point $(3 \cos \theta, 2 \sin \theta)$ the equation of the tangent is:

$$\dfrac{(y - 2 \sin \theta)}{(x - 3 \cos \theta)} = \dfrac{2 \cos \theta}{-3 \sin \theta}$$

$-3 \sin \theta \, (y - 2 \sin \theta) = 2 \cos \theta \, (x - 3 \cos \theta)$

$-3y \sin \theta + 6 \sin^2 \theta = 2x \cos \theta - 6 \cos^2 \theta$

$6 \, (\sin^2 \theta + \cos^2 \theta) = 2x \cos \theta + 3y \sin \theta$

$2x \cos \theta + 3y \sin \theta = 6$ (since $\sin^2 \theta + \cos^2 \theta = 1$)

(c) At (3, 0), $\theta = 0$. The equation is $x = 3$.

4 (a) $x = \dfrac{1}{2} t \Rightarrow t = 2x$

So, $y = 3(2x)^2 - 5 \times 2x = 12x^2 - 10x$

(b) $x = 5t \Rightarrow t = 0.2x$

So, $y = 6 - \dfrac{10}{0.2x} = 6 - \dfrac{50}{x}$

5 $x = \ln 2t \Rightarrow \dfrac{dx}{dt} = \dfrac{1}{t}$

$y = 2t + 3 \ln t \Rightarrow \dfrac{dy}{dt} = 2 + \dfrac{3}{t}$

$\dfrac{dy}{dx} = \dfrac{2 + 3/t}{1/t} = t\,(2 + \dfrac{3}{t}) = 2t + 3$

At $t = 1$, $x = 0.693$ (to 3 s.f.), $y = 2$, $\dfrac{dy}{dx} = 5$

The equation of the tangent is:

$\dfrac{y - 2}{x - 0.693} = 5$

$(y - 2) = 5(x - 0.693)$

$y \approx 5x - 1.47$

6 (a) (30, 100)

(b) $\sqrt{(9^2 + 2.7^2)} = 9.4$ m (to 2 s.f.)

(c) Velocity $= \begin{bmatrix} 3 \\ 1.2 \end{bmatrix}$ m s^{-1}

(d) Velocity $= \begin{bmatrix} 3 \\ 0.3 \end{bmatrix}$ m s^{-1}

So the speed is $\sqrt{(3^2 + 0.3^2)} = 3.01$ m s^{-1}.

Product rule

1 (a) $\dfrac{\cos x}{\sin x} = \cot x$ (b) $-3 \cos^2 x \sin x$

(c) $-3x^2 \sin x^3$ (d) $\dfrac{-8}{(4x - 3)^3}$

(e) $8x + 6xy + 3x^2 \dfrac{dy}{dx} - 2y \dfrac{dy}{dx} = 0$

$\dfrac{dy}{dx} = \dfrac{8x + 6xy}{2y - 3x^2}$

(f) $\dfrac{dx}{dt} = 3$, $\dfrac{dy}{dt} = 8t$ so $\dfrac{dy}{dx} = \dfrac{8t}{3} = 2\dfrac{2}{3}t$

(g) $2xe^{3x} + 3x^2e^{3x} = xe^{3x}(2 + 3x)$

(h) $\dfrac{2(4x+5) - 4(2x-3)}{(4x+5)^2} = \dfrac{22}{(4x+5)^2}$

(i) $\dfrac{(0 \times \cos x) - (1 \times -\sin x)}{\cos^2 x} = \dfrac{\sin x}{\cos^2 x} = \sec x \tan x$

(j) $8x - 5 - 4y\dfrac{dy}{dx} + 8\dfrac{dy}{dx} = 0 \Rightarrow \dfrac{dy}{dx} = \dfrac{8x-5}{4y-8}$

(k) $4\cos 3x \cos 2x - 6\sin 2x \sin 3x$

(l) $\dfrac{dx}{d\theta} = -2\sin\theta, \quad \dfrac{dy}{d\theta} = 8\cos 2\theta$

$\dfrac{dy}{dx} = \dfrac{8\cos 2\theta}{-2\sin\theta} = \dfrac{-4\cos 2\theta}{\sin\theta}$

2 $y = \tan x = \dfrac{\sin x}{\cos x}$

$\dfrac{dy}{dx} = \dfrac{\cos x \cos x - \sin x (-\sin x)}{\cos^2 x} = \dfrac{1}{\cos^2 x} = \sec^2 x$

3 $\dfrac{dy}{dx} = 0.5\sin x\, e^{0.5x} + \cos x\, e^{0.5x}$

$= e^{0.5x}(0.5\sin x + \cos x)$

At a turning point, $0.5\sin x + \cos x = 0$, so $\tan x = -2$

$x = 2.03, 5.18$ (to 3.s.f) (since $0 \le x \le 6$)

Turning points are $(2.03, 2.47)$ and $(5.18, -11.9)$.

$x = 2, \quad \dfrac{dy}{dx} = 0.105$

$x = 2.1, \quad \dfrac{dy}{dx} = -0.209$ $\Big\}$ $(2.03, 2.47)$ is a maximum

$x = 5.1, \quad \dfrac{dy}{dx} = -1.09$

$x = 5.2, \quad \dfrac{dy}{dx} = 0.36$ $\Big\}$ $(5.18, -11.9)$ is a minimum

4 $\dfrac{dy}{dx} = \dfrac{(3x+2)^3 - 9x(3x+2)^2}{(3x+2)^6} = \dfrac{(3x+2)^2(3x+2-9x)}{(3x+2)^6}$

$= \dfrac{(3x+2)^2(2-6x)}{(3x+2)^6}$

At the maximum, $(3x+2)^2(2-6x) = 0$. So the positive value of x which gives a maximum is $\dfrac{1}{3}$.

5 (a) $3x^2 - 3y^2\dfrac{dy}{dx} + 3 + 2\dfrac{dy}{dx} = 0 \Rightarrow \dfrac{dy}{dx} = \dfrac{3x^2+3}{3y^2-2}$

(b) At $(2, 3)$, the gradient is $\dfrac{15}{25} = 0.6$

Equation of the tangent is $\dfrac{y-3}{x-2} = 0.6$

$y - 3 = 0.6(x - 2) \Rightarrow y = 0.6x + 1.8$

6 (a) $f'(x) = \dfrac{(\cos x + 3)\cos x + \sin x(\sin x + 1)}{(\cos x + 3)^2} = \dfrac{3\cos x + \sin x + 1}{(\cos x + 3)^2}$

(b) $f(0) = 0.25$ and $f'(0) = 0.25$

The equation of the tangent is $y = 0.25x + 0.25$.

Volume

1 (a) (i) $y = \sin(x^3) \Rightarrow \dfrac{dy}{dx} = 3x^2 \cos(x^3)$

(ii) $y = \sin^3 x \Rightarrow \dfrac{dy}{dx} = 3\sin^2 x \cos x$

(b) From (i) $\displaystyle\int 3x^2 \cos(x^3)\,dx = \sin(x^3) + c$

So $\displaystyle\int x^2 \cos(x^3)\,dx = \dfrac{1}{3}\sin^3 x + k$

From (ii) $\displaystyle\int 3\sin^2 x \cos x\,dx = \sin^3 x + c$

So $\displaystyle\int \sin^2 x \cos x\,dx = \dfrac{1}{3}\sin^3 x + k$

2 (a) $\displaystyle\int \cos 2\theta\,d\theta = \dfrac{1}{2}\sin 2\theta + c$

(b) $\cos 2\theta = 1 - 2\sin^2\theta \Rightarrow \sin^2\theta = \dfrac{1}{2}(1 - \cos 2\theta)$

$\displaystyle\int \sin^2\theta\,d\theta = \dfrac{1}{2}\int(1 - \cos 2\theta)\,d\theta$

$= \dfrac{1}{2}\theta - \dfrac{1}{4}\sin 2\theta + c$

(c) $\cos 2\theta = 2\cos^2\theta - 1 \Rightarrow \cos^2\theta = \dfrac{1}{2}(\cos 2\theta + 1)$

$\displaystyle\int_0^{\pi/2} 0.5\cos^2\theta\,d\theta = \int_0^{\pi/2}(\cos 2\theta + 1)\,d\theta$

$= \dfrac{1}{4}\left[\dfrac{1}{2}\sin 2\theta + \theta\right]_0^{\pi/2}$

$= 0.393$ (to 3 s.f.)

3 (a) $\sin 2x \cos 4x = \dfrac{1}{2}(\sin 6x + \sin(-2x))$

$\displaystyle\int_{0.5}^1 \sin 2x \cos 4x\,dx = \dfrac{1}{2}\int_{0.5}^1(\sin 6x - \sin 2x)\,dx$

$= \dfrac{1}{2}\left[-\dfrac{1}{6}\cos 6x + \dfrac{1}{2}\cos 2x\right]_{0.5}^1$

$= -0.402$ (to 3 s.f.)

(b) $\displaystyle\int_1^{1.5} \cos x \cos 3x\,dx = \dfrac{1}{2}\int_1^{1.5}(\cos 4x + \cos 2x)\,dx$

$= \dfrac{1}{2}\left[\dfrac{1}{4}\sin 4x + \dfrac{1}{2}\sin 2x\right]_1^{1.5}$

$= -0.132$ (to 3 s.f.)

4 Volume $= \pi\displaystyle\int_0^{\pi/2} y^2\,dx = \pi\int_0^{\pi/2} \sin^2 x\,dx$

$= \dfrac{1}{2}\pi\displaystyle\int_0^{\pi/2}(1 - \cos 2x)\,dx$

$= \dfrac{1}{2}\pi\left[x - \dfrac{1}{2}\sin 2x\right]_0^{\pi/2} = 2.47$ (to 3.s.f.)

5 (a) $\frac{1}{3}e^{x^3} + c$ (b) $\frac{1}{91}(7x+12)^{13} + c$

    (c) $-\frac{1}{5}e^{-5x} + c$ (d) $\frac{1}{5}(x^2+3x)^5 + c$

    (e) $\ln|x^2+3x| + c$ (f) $-\frac{1}{6}\cos 6x + c$

    (g) $\frac{1}{6}\sin(3x^2+4) + c$

6 Volume $= \pi \int_0^3 x^2\, dy = \pi \int_0^3 (y+1)^2\, dy$

    $= \pi \int_0^3 (y^2+2y+1)\, dy = \pi \left[\frac{y^3}{3} + y^2 + y \right]_0^3$

    $= 66$ (to 2 s.f.)

Integration techniques

1 (+ c is omitted from all integrals)

    (a) $-\frac{1}{5}\cos 5x$ (b) $-\frac{x}{5}\cos 5x + \frac{1}{25}\sin 5x$

    (c) $\frac{3}{5}x^5 - \frac{1}{6}x^6$ (d) $\frac{1}{3}\ln|x^3-3|$

    (e) $\frac{1}{2}e^{x^2}$ (f) $\frac{1}{5}\ln|5x+2|$

    (g) $\frac{1}{3}xe^{3x} - \frac{1}{9}e^{3x}$ (h) $\frac{1}{4}\tan(4x-3)$

    (i) $\frac{1}{2}x + \frac{1}{12}\sin 6x$ (j) $-3\ln|2-x|$

    (k) $\frac{4}{3}x^{3/2} - 6x^{1/2}$ (l) $\frac{1}{55}(5x-3)^{11}$

    (m) $\sqrt{(x^2+4)}$ (n) $\sqrt{(2x+4)}$

    (o) $-\frac{1}{8}e^{8x}$ (p) $\frac{1}{2}x^2 - 3\ln|x|$

    (q) $\frac{1}{10}x(3+2x)^5 - \frac{1}{120}(3+2x)^6$

    (r) $\frac{1}{12}(5-\cos 3x)^4$ (s) $\frac{1}{4}x^4\ln|x| - \frac{1}{16}x^4$

    (t) $-x^2\cos x + 2x\sin x + 2\cos x$

2 (a) $\frac{1}{6}e^{2x} + c$

    (b) $\int 2e^{-3x}\, dx = -\frac{2}{3}e^{-3x} + c$

3 $\left[x \times \frac{1}{6}(x+3)^6 \right]_0^1 - \int_0^1 \frac{1}{6}(x+3)^6\, dx$

    $= \frac{1}{6} \times 4^6 - \left[\frac{1}{42}(x+3)^7 \right]_0^1$

    ≈ 344.6

4 At $x = 1$, $\frac{1}{x^2} = 1 = \frac{1}{4x-3}$

    At $x = 3$, $\frac{1}{x^2} = \frac{1}{9} = \frac{1}{4x-3}$

$\int_1^3 \frac{1}{x^2}\, dx = \left[-\frac{1}{x} \right]_1^3 = \frac{2}{3}$

$\int_1^3 \frac{1}{4x-3}\, dx = \left[\frac{1}{4}\ln|4x-3| \right]_1^3 \approx 0.5493$

Area enclosed ≈ 0.117

5 $\frac{d}{dx}\left[\sin(3x^2+4) \right] = 6x\cos(3x^2+4)$

    $\left[\frac{1}{6}\sin(3x^2+4) \right]_0^6 \approx -0.022$

6 $x = 4\sin u \Rightarrow dx = 4\cos u\, du$

    $\int \frac{4\cos u}{4\sqrt{(1-\sin^2 u)}}\, du = u + c = \sin^{-1}(\frac{1}{4}x) + c$

7 (a) $5 + \frac{5x+1}{x^2-x-12}$

    $= 5 + \frac{2}{x+3} + \frac{3}{x-4}$

    (b) $\left[5x + 2\ln|x+3| + 3\ln|x-4| \right]_0^1$

    $= 5 + 2\ln\frac{4}{3} + 3\ln\frac{3}{4}$

    ≈ 4.712

Polynomial approximations/First principles

1 $x_1 = 1.384$; $x_2 = 1.286$; $x_3 = 1.281 = x_4$

2 $y = 2e^2 x - e^2$

3 $\cos x \approx 1 - \frac{1}{2}x^2 + \frac{1}{24}x^4 - \dots$

    $\Rightarrow \cos 3x \approx 1 - \frac{9}{2}x^2 + \frac{27}{8}x^4 - \dots$

4 (a) Area ACB $= \frac{1}{2}r^2\theta - \frac{1}{2}r^2\sin\theta$

    So $\frac{1}{2}r^2\theta - \frac{1}{2}r^2\sin\theta = \frac{1}{10}\pi r^2$

    $\Rightarrow \theta - \frac{1}{5}\pi = \sin\theta$

    (b)

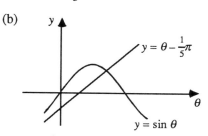

    (c) $1 + \frac{1}{5}\pi \approx 1.63$

5 (a) $f(x) = \tan x$, $f'(x) = \frac{1}{\cos^2 x}$,

    $f''(x) = \frac{2\sin x}{\cos^3 x}$, $f'''(x) = \frac{2\cos^4 x + 6\sin^2 x\cos^2 x}{\cos^6 x}$

$f(0) = 0, f'(0) = (1), f''(0) = 0, f'''(0) = 2$

$f(x) \approx x + \dfrac{1}{3} x^3$

(b) $\dfrac{\tan 0.5 - (0.5 + \frac{1}{3} 0.5^3)}{\tan 0.5} \times 100 \approx 0.8\%$

6   $\sqrt{(1 - e^{-h})} \approx \sqrt{(1 - (1 - h))} = \sqrt{h}$

(a)   LHS ≈ 0.308, RHS ≈ 0.316

(b)   (i) $e^{-x} \le 1 \Rightarrow x \ge 0$ (ii) $0 \le f(x) < 1$

(c)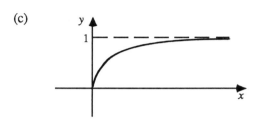

MISCELLANEOUS EXERCISES

FOUNDATIONS

1   (a) $p = 3, q = -4$

(b) $(x + 3)^2 - 4 = x^2 + 6x + 5$
$b = 6, c = 5$

(c) $x^2 + 6x + 5 = (x + 1)(x + 5)$
$k = 1, h = 5$

(d) $(-1, 0)$ and $(-5, 0)$

(e) $(0, 5)$

2   (a) 0

(b) $(x + 3)(x^2 + 3x + 2) = (x + 1)(x + 2)(x + 3)$

(c) $x^3 - x$

(d) The graph of $y = g(x - 2)$ is the same as the graph of $y = g(x)$ translated by $+2$ in the x-direction.

3   (a) Let £a be the fixed costs.
Let £$\dfrac{b}{V}$ be the cost of wages.
Let £cV^2 be the cost of fuel
Total cost $= £\left(a + \dfrac{b}{V} + cV^2 \right)$

(b)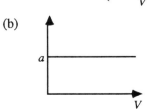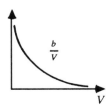

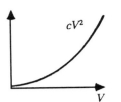

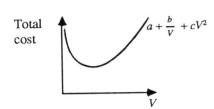

4   (a) 4096, the 7th term.

(b) 3003, the 77th triangular number.

5   $\dfrac{£200 \left(1 - 1.115^{12}\right)}{1 - 1.115} = £4682.28$

6   (a) £25 736.80

(b) £21 048$(1.105^n - 1)$

(c) 19 years

7   (a) $x(x - 3) = 20 \Rightarrow x^2 - 3x - 20 = 0$

(b) $x = \dfrac{3}{2} \pm \dfrac{\sqrt{89}}{2} = \dfrac{1}{2} (3 \pm \sqrt{89})$
The numbers are $\dfrac{1}{2} (3 + \sqrt{89})$ and $\dfrac{1}{2} (-3 + \sqrt{89})$

8   $10 + 10 \left(\dfrac{7}{10}\right) + 10 \left(\dfrac{7}{10}\right)^2 + \dots$
$= \dfrac{10}{1 - \frac{7}{10}} = 33\dfrac{1}{3}$
He does not reach his intended destination.

9   $0.513 \left(1 + \dfrac{1}{1000} + \left(\dfrac{1}{1000}\right)^2 + \dots\right)$
$= \dfrac{0.513}{0.999} = \dfrac{19}{37}$

10 (a) $u_1 = \dfrac{1}{4}, u_2 = \dfrac{5}{16}, u_3 = \dfrac{21}{64}$

(b) $u_n = \dfrac{1}{4} (u_{n-1} + 1)$

Substituting the formula for u_{n-1} and u_n,

RHS $= \dfrac{1}{4} \left(\dfrac{1}{3} - \dfrac{1}{3} \left(\dfrac{1}{4}\right)^{n-1} + 1 \right)$

$= \dfrac{1}{4} \left(\dfrac{4}{3} - \dfrac{1}{3} \left(\dfrac{1}{4}\right)^{n-1} \right)$

$= \dfrac{1}{3} - \dfrac{1}{3} \left(\dfrac{1}{4}\right)^n$

$=$ LHS

INTRODUCTORY CALCULUS

1 When $y = 1.4$, $x = \pm 0.5$

$$2 \int_0^{0.5} (0.4 + 4x^2)dx + 2 \times 0.05 \times 1.4 = 0.8733$$

Volume $= 3 \times 0.8733 \approx 2.62$ m³

2. (a) $\tan \alpha = 1$ and $\sec^2 \alpha = 2$, so

$$y = x - \frac{10x^2}{\frac{u^2}{2}}$$

$$\Rightarrow y = x - \frac{x^2}{90}$$

(b) y max occurs when $\frac{dy}{dx} = 0$ i.e. when $x = 45$.
y max $= 22.5$ metres

(c) If $x = 30$, then $y = 20$.

(d) When $y = 0$, $x = 0$ or 90.
90 metres from 0.

(e)

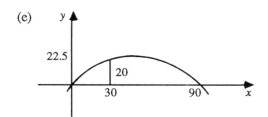

3

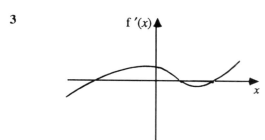

4 (a) $\frac{dy}{dx} = 12x^3 - 12x^2$

(b) $\frac{dy}{dx} = 0$ when $12x^2 (x - 1) = 0$

i.e. $x = 0$ or 1.

The stationary points are $(0, 0)$ and $(1, -1)$.

(c)

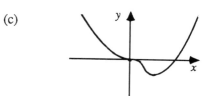

5 $\frac{dy}{dx} = 6x - 1$

$\frac{dy}{dx} = 11$ at $(2, 8)$

$y = 11x - 14$

6 (a) $x^3 - 4x = 0$ when $x = -2, 0$ or 2.

$$\int_0^2 (x^3 - 4x)dx = \left[\frac{1}{4}x^4 - 2x^2 \right]_0^2 = -4$$

The area is 4 m².

(b) $\frac{dy}{dx} = 0$ when $3x^2 = 4 \Rightarrow x \approx \pm 1.1547$

The highest point is at

$(-1.1547)^3 - 4(-1.1547) \approx 3.079$

3.08 metres

7 (a)

0	0.25	0.5	0.75	1
2	1.882	1.6	1.28	1

$\frac{1}{2} \times \frac{1}{4} [2 + 2 (1.882 + 1.6 + 1.28) + 1]$

≈ 1.5655

(b) $\frac{2}{1+x^2}$ is an even function and so the area is

$2 \times 1.5655 \approx 3.131$

8 (a) $\int (x^2 + x - 2)dx = \frac{1}{3} x^3 + \frac{1}{2} x^2 - 2x + c$

$$\left[\frac{1}{3}x^3 + \frac{1}{2} x^2 - 2x + c \right]_0^1 = -1\frac{1}{6}$$

$$\left[\frac{1}{3}x^3 + \frac{1}{2} x^2 - 2x + c \right]_1^2 = 1\frac{5}{6}$$

(b) $1\frac{1}{6} + 1\frac{5}{6} = 3$

9 $\frac{2^{3.001} - 2^3}{0.001} \approx 5.55$

FUNCTIONS

1 (a) $a = 5$ and $b = 3$

(b) $5 + 3 \sin \pi t = 6$
$3 \sin \pi t = 1$
$\pi t = \sin^{-1} \frac{1}{3} = 0.3398$
$t \approx 0.108$ minutes i.e. 6.5 seconds.

2 (a) $\cos (\frac{1}{2} \theta + 10°) = 0.75$

$\frac{1}{2} \theta + 10° = 41.4°$ or $318.6°$

For $0° \leq \theta \leq 360°$, $0° \leq \frac{1}{2}\theta \leq 180°$, so the only

solution comes from:

$\frac{1}{2}\theta + 10° = 41.4°$, which gives $\theta = 62.8°$

(b) $\sin \theta = \frac{1}{3}$ gives $\theta = 19.5°, 160.5°$

$\sin \theta = -\frac{1}{3}$ gives $\theta = 199.5°, 340.5°$

So the solutions are $19.5°, 160.5°, 199.5°, 340.5°$

3 (a) $4 = 2.3 \cos (0.5t) + 6.1 \Rightarrow -2.1 = 2.3 \cos (0.5t)$

$0.5t = \cos^{-1} \left(\frac{-2.1}{2.3} \right)$

$t = 5.443$, so the time is 5.27 p.m.

(b) $\frac{dD}{dt} = -1.15 \sin (0.5t)$, so the greatest rate at which the tide falls is 1.15 metres per hour.

4 (a) $20\theta = 30 \Rightarrow \theta = 1.5$

(b) Area of sector ACB $= \frac{1}{2} \times 20^2 \times 1.5$ m^2

Area of triangle ACB $= \frac{1}{2} \times 20^2 \times \sin 1.5$ m^2

Shaded area $= 100.5$ m^2 (to 1 d.p.)
$100.5 \times 1.2 = 120.6$, so there are 120 children in the party.

5 (a) The annual interest rate is 10%.

(b) $1.1^n = \frac{T}{1000}$

$\log_{10} 1.1^n = \log_{10} T - \log_{10} 1000 = \log_{10} T - 3$

$n \log_{10} 1.1 = \log_{10} T - 3 \Rightarrow n = \frac{\log_{10} T - 3}{\log_{10} 1.1}$

(c) The money has doubled when $T = 2000$.
This gives $n = 7.27$, so it will have doubled after 8 years.

6 (a) $a = 30, b = 20, c = \frac{1}{3}$

(b) $45 = 30 + 20 \cos (180t + 60°)$

$\cos (180t + 60°) = 0.75$

$180t + 60 = 41.4, 318.6, 401.4, 678.6, 761.4$

$t = 1.4, 1.9, 3.4, 3.9$ seconds

7 $\frac{dy}{dx} = ke^{kx}$

At $(\frac{1}{k}, e)$, $\frac{dy}{dx} = ke$

$y = kex$

8 (a)

Cross-sectional area $= (4 + 1.5 \sin \theta) \, 1.5 \cos \theta$
Depth $= 2$
Volume $= (4 + 1.5 \sin \theta) \, 3 \cos \theta$

$= 12 \cos \theta + \frac{9}{4} \sin 2\theta$

(b) By plotting V against θ and tracing along the graph, the maximum is found to occur when $\theta = 0.31$.

$V_{max} \approx 12.7$ m^3

MATHEMATICAL METHODS

1 (a) $\cos t = \frac{1}{5}(x - 3), \sin t = \frac{1}{5}(y - 4)$

(b) $\frac{1}{25}(x - 3)^2 + \frac{1}{25}(y - 4)^2 = 1$

(c) Rearranging the equation in (b) gives
$(x - 3)^2 + (y - 4)^2 = 5^2$, the equation of a circle, centre (3, 4) and radius 5.

2 (a) $x^2 = (\frac{1}{2}p)^2 + (\frac{1}{2}q)^2 - 2(\frac{1}{2}p)(\frac{1}{2}q) \cos \theta$

$\Rightarrow x^2 = \frac{1}{4}p^2 + \frac{1}{4}q^2 - \frac{1}{2}pq \cos \theta$

(b) $y^2 = (\frac{1}{2}p)^2 + (\frac{1}{2}q)^2 - 2(\frac{1}{2}p)(\frac{1}{2}q) \cos (180 - \theta)$

$\Rightarrow y^2 = \frac{1}{4}p^2 + \frac{1}{4}q^2 - \frac{1}{2}pq \cos (180 - \theta)$

(c) $-\cos \theta$ is a reflection of $\cos \theta$ in the x-axis and $\cos (180 - \theta)$ is a translation of $\cos \theta$ by $180°$ followed by a reflection in the y-axis. Both transformations have the same effect on the cosine graph.

(d) Replace $\cos (180 - \theta)$ by $-\cos \theta$ in y^2,
$x^2 + y^2 = \frac{1}{2}p^2 + \frac{1}{2}q^2 \Rightarrow 2(x^2 + y^2) = p^2 + q^2$

(e) Substituting $x = 3, y = 5$ and $p = 6$ into the result from (d):

$2(9 + 25) = 36 + q^2 \Rightarrow q = 5.7$ cm

3 (a) If $u = 2x, y = \text{soc } u$ then $\frac{du}{dx} = 2, \frac{dy}{du} = \sqrt{(1 + u^2)}$

$\frac{dy}{dx} = \frac{dy}{du} \times \frac{du}{dx} = 2\sqrt{(1 + u^2)} = 2\sqrt{(1 + 4x^2)}$

(b) $\frac{dy}{dx} = \frac{\sqrt{(1 + x)}}{2\sqrt{x}}$

4 $5 \cos \theta - 2 \sin \theta = r \cos (\theta + \alpha)$

where $r = \sqrt{(5^2 + 2^2)} = \sqrt{29}$ and $\alpha = \tan^{-1} \frac{2}{5} = 21.8°$

$\sqrt{29} \cos (\theta + 21.8) = 0.6$

$\Rightarrow \cos (\theta + 21.8) = 0.111$

$\Rightarrow \theta + 21.8 = 83.6, 276.4, \ldots$

$\theta = 61.8°$ and $254.6°$

5 If a common point exists, then

$$\begin{bmatrix} 0 \\ 2 \\ -3 \end{bmatrix} + s \begin{bmatrix} 1 \\ -1 \\ -1 \end{bmatrix} = \begin{bmatrix} -1 \\ 6 \\ -1 \end{bmatrix} + t \begin{bmatrix} 2 \\ 1 \\ -1 \end{bmatrix}$$

and

$$s = -1 + 2t \quad \text{①}$$
$$2 - s = 6 + t \quad \text{②}$$
$$-3 - s = -1 - t \quad \text{③}$$

From ① and ②, $s = -3$, $t = -1$. These values of s and t satisfy ③ and so a common point, $(-3, 5, 0)$, exists.

$$\mathbf{r} = \begin{bmatrix} 0 \\ 2 \\ -3 \end{bmatrix} + \lambda \begin{bmatrix} 1 \\ -1 \\ -1 \end{bmatrix} + \mu \begin{bmatrix} 2 \\ 1 \\ -1 \end{bmatrix}$$

6 $1 + \frac{1}{2}h - \frac{1}{8}h^2$

$$\int_0^{0.04} \sqrt{(1 + \sqrt{x})}\, dx \approx \int_0^{0.04} (1 + \frac{1}{2}x^{1/2} - \frac{1}{8}x)\, dx$$

$$\approx 0.0426$$

7 (a) $\begin{bmatrix} -5 \\ -3 \\ -1 \end{bmatrix} \cdot \begin{bmatrix} 3 \\ -5 \\ -1 \end{bmatrix} = \sqrt{35}\,\sqrt{35}\cos A \Rightarrow \cos A = \frac{1}{35}$

$A \approx 88.4°$

(b) $\mathbf{r} = \begin{bmatrix} 30 \\ 50 \\ 80 \end{bmatrix} + \lambda \begin{bmatrix} 5 \\ 3 \\ 1 \end{bmatrix} + \mu \begin{bmatrix} 3 \\ -5 \\ -1 \end{bmatrix}$

(c) To get to the top right of the V,

$$\begin{bmatrix} 30 \\ 50 \\ 80 \end{bmatrix} + \begin{bmatrix} 5 \\ 3 \\ 1 \end{bmatrix} + \begin{bmatrix} 3 \\ -5 \\ -1 \end{bmatrix} = \begin{bmatrix} 38 \\ 48 \\ 80 \end{bmatrix}$$

So $\begin{bmatrix} -38 \\ -48 \\ -80 \end{bmatrix}$ will take the arm back.

8 (a) $\frac{dy}{dt} = ky(N - y)$

(b) $N = 500$ and when $t = 1$, $y = 6$ and $\frac{dy}{dt} = 5$.

$5 = k \times 494 \times 6 \Rightarrow k = 0.0017$

$\frac{dy}{dt} = 0.0017y\,(500 - y)$

(c)

t	y	$\frac{dy}{dt}$	dt	dy	$t + dt$	$y + dy$
1	6	5	1	5	2	11
2	11	9.1	1	9.1	3	20.1

Approximately 20 will be infected after 3 weeks.

9 (a) 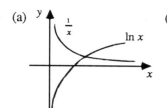 (b)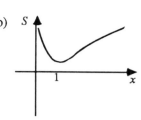

(c) $\frac{dS}{dx} = -\frac{1}{x^2} + \frac{1}{x} = 0 \Rightarrow = 1$

The minimum value is $S = 1$

CALCULUS METHODS

1 (a) $\int (2x^{-2} - x^{-3})\, dx = c - \frac{2}{x} + \frac{1}{2x^2}$

(b) $\int_1^{10} (\frac{1}{s^2} - 2 + s^2)\, ds = \left[-\frac{1}{s} - 2s + \frac{1}{3}s^3\right]_1^{10} = 315.9$

(c) $\int_1^4 x^{-\frac{1}{2}}\, dx = \left[2x^{\frac{1}{2}}\right]_1^4 = 2$

2 $\frac{dy}{dx} = 0$ when $\dfrac{(3x+2)^2 - x \times 2 \times 3\,(3x+2)}{(3x+2)^4} = 0$

$(3x + 2)^2 = 6x(3x + 2) \Rightarrow x = -\frac{2}{3}$ or $\frac{2}{3}$

$x = \frac{2}{3}$

3 $\frac{dy}{dx} = 2x \ln x + x$

At (e, e^2), $\frac{dy}{dx} = 3e$

$y = 3ex - 2e^2$

4 (a) $1500 = \pi r^2 h \Rightarrow h = \dfrac{1500}{\pi r^2}$

(b) $A = 2\pi rh + 2\pi r^2$

$= \dfrac{3000}{r} + 2\pi r^2$

(c) $L = k\left(\dfrac{3000}{r} + 2\pi r^2\right)$

$\dfrac{dL}{dr} = 0$ when $-\dfrac{3000}{r^2} + 4\pi r = 0$

$r = 6.2$ cm, $h = 12.4$ cm

(d) Sphere: 633.7 cm²   Cylinder: 725.4 cm²

The spherical teapot is better.

5 (a) $\mathbf{v} = \begin{bmatrix} 20 \\ 12 - 10t \end{bmatrix}$

(b) Speed² $= 20^2 + (12 - 10t)^2$

Speed $= \sqrt{(544 - 240t + 100t^2)}$

The speed is least when $12 - 10t = 0 \Rightarrow t = 1.2$

The least speed is 20 ms⁻¹

6 (a) The volume can be split into discs of radius r and thickness dy where each disc has volume

$$\pi r^2 \, dy = \pi \, (y \tan 30°)^2 \, dy$$

Therefore, $V = \pi \displaystyle\int_0^h (y \tan 30°)^2 \, dy$

(b) $V = \pi (\tan 30°)^2 \displaystyle\int_0^h y^2 dy$

$$= \frac{1}{3} \pi \left[\frac{1}{3} y^3 \right]_0^h$$

$$= \frac{1}{9} \pi h^3$$

7 (a) $2(x - y) \left(1 - \dfrac{dy}{dx}\right) = 1 + \dfrac{dy}{dx}$

$$\frac{dy}{dx} = \frac{2x - 2y - 1}{2x - 2y + 1}$$

(b) -1 at $(0, 0)$

   3 at $(0, 1)$

   $\dfrac{1}{3}$ at $(1, 0)$

(c)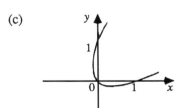

8 $\pi \displaystyle\int_0^2 \frac{1}{(2x+1)^2} \, dx = \left[-\frac{\pi}{2\,(2x+1)} \right]_0^2 = \frac{2}{5} \pi$

9 $\dfrac{1}{3} \displaystyle\int_2^{26} \frac{u-2}{u^2} \frac{1}{3} \, du = \frac{1}{9} \int_2^{26} \left(u^{-1} - 2u^{-2} \right) du$

$$= \frac{1}{9} \left[\ln u + 2u^{-1} \right]_2^{26}$$

$$\approx 0.182$$

10 (a) The angle of the sector is $\sin^{-1}\left(\dfrac{1}{2}\right) = 30°$

   The area is therefore $\dfrac{30}{360} \times \pi \times 1^2 = \dfrac{1}{12} \pi$

(b) $\dfrac{1}{12} \pi = \displaystyle\int_0^{1/2} \sqrt{(1 - x^2)} \, dx - \frac{1}{2} \times \frac{1}{2} \times \frac{\sqrt{3}}{2}$

$$\Rightarrow \pi = 12 \int_0^{1/2} \sqrt{(1 - x^2)} \, dx - \frac{1}{2} \sqrt{27}$$

(c) $\sqrt{(1 + z)} \approx 1 + \dfrac{1}{2} z - \dfrac{1}{8} z^2$

   $\sqrt{(1 - x^2)} \approx 1 - \dfrac{1}{2} x^2 - \dfrac{1}{8} x^4$

(d) $\displaystyle\int_0^{1/2} \sqrt{(1 - x^2)} \, dx \approx \left[x - \frac{1}{6} x^3 - \frac{1}{40} x^5 \right]_0^{1/2}$

$$\approx 0.4784$$

$$\sqrt{(25 + 2)} = 5 \sqrt{\left(1 + \frac{2}{25}\right)}$$

$$\approx 5 \left(1 + \frac{1}{25} - \frac{1}{8} \left(\frac{2}{25} \right)^2 \right)$$

$$\approx 5.196$$

(e) $\pi \approx 12 \times 0.4784 - \dfrac{1}{2} \times 5.196$

$$\approx 3.14$$

INDEX